Global Issues Surrounding Outer Space Law and Policy

Doo Hwan Kim
Korea Society of Air and Space Law and Policy, South Korea

A volume in the Advances in Public Policy and Administration (APPA) Book Series

IGI Global
PUBLISHER of TIMELY KNOWLEDGE

Published in the United States of America by
IGI Global
Information Science Reference (an imprint of IGI Global)
701 E. Chocolate Avenue
Hershey PA, USA 17033
Tel: 717-533-8845
Fax: 717-533-8661
E-mail: cust@igi-global.com
Web site: http://www.igi-global.com

Library of Congress Cataloging-in-Publication Data

Names: Kim, Tu-hwan, 1934- author.
Title: Global issues surrounding outer space law and policy / by Doo Hwan Kim.
Description: Hershey, PA : Information Science Reference, 2021. | Includes bibliographical references and index. | Summary: "This authored book addresses the global issues and solutions pertaining to outer space law and policy, addressing the United Nations international five space treaties, the main contents of thirty countries' domestic space law and policy as well as their national space agencies and two space international organizations"-- Provided by publisher.
Identifiers: LCCN 2021010769 (print) | LCCN 2021010770 (ebook) | ISBN 9781799874072 (hardcover) | ISBN 9781799874089 (paperback) | ISBN 9781799874096 (ebook)
Subjects: LCSH: Space law. | Outer space--Government policy. | United Nations. General Assembly. Committee on the Peaceful Uses of Outer Space. | International Law Association.
Classification: LCC KZD1145 .K56 2021 (print) | LCC KZD1145 (ebook) | DDC 341.4/7--dc23
LC record available at https://lccn.loc.gov/2021010769
LC ebook record available at https://lccn.loc.gov/2021010770

This book is published in the IGI Global book series Advances in Public Policy and Administration (APPA) (ISSN: 2475-6644; eISSN: 2475-6652)

British Cataloguing in Publication Data
A Cataloguing in Publication record for this book is available from the British Library.

For electronic access to this publication, please contact: eresources@igi-global.com.

Advances in Public Policy and Administration (APPA) Book Series

ISSN:2475-6644
EISSN:2475-6652

Mission

Proper management of the public sphere is necessary in order to maintain order in modern society. Research developments in the field of public policy and administration can assist in uncovering the latest tools, practices, and methodologies for governing societies around the world.

The **Advances in Public Policy and Administration (APPA) Book Series** aims to publish scholarly publications focused on topics pertaining to the governance of the public domain. APPA's focus on timely topics relating to government, public funding, politics, public safety, policy, and law enforcement is particularly relevant to academicians, government officials, and upper-level students seeking the most up-to-date research in their field.

Coverage

- Government
- Law Enforcement
- Political Economy
- Politics
- Public Administration
- Public Funding
- Public Policy
- Resource Allocation
- Urban Planning

IGI Global is currently accepting manuscripts for publication within this series. To submit a proposal for a volume in this series, please contact our Acquisition Editors at Acquisitions@igi-global.com or visit: http://www.igi-global.com/publish/.

The Advances in Public Policy and Administration (APPA) Book Series (ISSN 2475-6644) is published by IGI Global, 701 E. Chocolate Avenue, Hershey, PA 17033-1240, USA, www.igi-global.com. This series is composed of titles available for purchase individually; each title is edited to be contextually exclusive from any other title within the series. For pricing and ordering information please visit http://www.igi-global.com/book-series/advances-public-policy-administration/97862. Postmaster: Send all address changes to above address.

Titles in this Series

For a list of additional titles in this series, please visit: www.igi-global.com/book-series

Global Politics, Political Participation, and the Rise of Nationalism Emerging Research and Opportunities
Emily Stacey (Rose State College, USA)
Information Science Reference • © 2021 • 213pp • H/C (ISBN: 9781799873433) • US $175.00

Redefining Theory and Practice to Guide Social Transformation Emerging Research and Opportunities
Beth Fisher-Yoshida (Columbia University, USA) and Joan Camilo Lopez (Columbia University, USA)
Information Science Reference • © 2021 • 200pp • H/C (ISBN: 9781799866275) • US $195.00

Africa's Platforms and the Evolving Sharing Economy
Immanuel Ovemeso Umukoro (Pan-Atlantic University, Nigeria) and Raymond Okwudiri Onuoha (Pan-Atlantic University, Nigeria)
Information Science Reference • © 2021 • 288pp • H/C (ISBN: 9781799832348) • US $195.00

Processual Perspectives on the Co-Production Turn in Public Sector Organizations
Anja Overgaard Thomassen (Aalborg University, Denmark) and Julie Borup Jensen (Aalborg University, Denmark)
Information Science Reference • © 2021 • 340pp • H/C (ISBN: 9781799849759) • US $195.00

Handbook of Research on Global Challenges for Improving Public Services and Government Operations
Cenay BABAOĞLU (Selçuk University, Turkey) Elvettin AKMAN (Süleyman Demirel University, Turkey) and Onur KULAÇ (Pamukkale University, Turkey)
Information Science Reference • © 2021 • 622pp • H/C (ISBN: 9781799849780) • US $295.00

Behavioral-Based Interventions for Improving Public Policies
Viorel Mihaila (University of Bucharest, Romania)
Information Science Reference • © 2021 • 318pp • H/C (ISBN: 9781799827313) • US $195.00

Examining International Land Use Policies, Changes, and Conflicts
G. N. Tanjina Hasnat (University of Chittagong, Bangladesh) and Mohammed Kamal Hossain (University of Chittagong, Bangladesh)
Information Science Reference • © 2021 • 439pp • H/C (ISBN: 9781799843726) • US $195.00

Effects of E-Authoritarianism and E-Protest on Contemporary Society
Nygmet Ibadildin (KIMEP University, Kazakhstan)
Information Science Reference • © 2020 • 300pp • H/C (ISBN: 9781799828631) • US $185.00

IGI Global
PUBLISHER of TIMELY KNOWLEDGE

701 East Chocolate Avenue, Hershey, PA 17033, USA
Tel: 717-533-8845 x100 • Fax: 717-533-8661
E-Mail: cust@igi-global.com • www.igi-global.com

Table of Contents

Preface

It is indeed a great pleasure for me to write this book containing the main contents and comments on the United Nations international five space treaties, the main contents of thirty countries' domestic space law and policy as well as their national space agencies and two space international organizations. As the outer space law has abundantly the density of the technical character, so it becomes possibly a shorter road to form a "global unification law (world law).

In order to understand easily the global issues and solutions on the outer space law and policy, I described only legal issues and the method of solution on the outer space law and policy.

Furthermore I am explained briefly the main contents, legal problems and solution of many international treaties, conventions, protocols, agreements etc. in the field of the outer space law as well as thirty countries' domestic space law of the United States, the United Kingdom, Russia Federation, China, Germany, France, Canada, the Netherlands, Japan, Australia, Spain, Brazil, India, the Republic of Korea, North Korea etc., European Space Agency and International space Station.

In addition, the structure and activities of the major international organizations related to the outer space laws and policies were introduced. Especially the major contents of the UN five international space treaties such as ① Space Treaty of 1967, ② Rescue Agreement of 1968, ③ Liability Convention of 1972, ④ Registration Convention of 1975, and ⑤Moon Agreement of 1979 were explained briefly, and their problems, solutions and comments are presented.

As the abovementioned UN five international space treaties had enacted more 40~50 years ago, because contents of the five space-related treaties does not fit the reality due to the rapid development of space science and technology, the changes in politics, economy and social structure and so it is necessary for us to revise the UN five international space treaties. The moon, Mars, asteroids, Jupiter, Saturn and Other Celestial Bodies of our solar system contained large quantity of natural resources.

Recently it is most severe competition among the space super-powers in order to mine and exploit the natural resources including Helium-3 from the moon so as to solve the serious problems of the future earth's energy crisis.

It is necessary and desirable for us the establishment of the new International Space Agency (ISA: tentative name) under the United Nations Committee on the Peaceful Use of Outer Space (UNCOPUOS) in order to be efficient and rapid exploitation and developments among the developed countries based on the Article 11, Paragraph 5 and Article 18 of the 1979 Moon Agreement. The establishment of a new ISA would lead to a strengthening of the international cooperation deemed essential by the global community towards joint undertakings and exploitation of the natural resources in the moon, Mars, asteroids, Jupiter, Saturn and Other Celestial Bodies.

The ISA is the international organization through which States manage and control the exploitation and developments of the natural resources in the moon, Mars, Venus, asteroid, Mercury, Jupiter, Saturn and Other Celestial Bodies.

It is indeed a great desirable and necessary for us to create a new Asian Space Agency (ASA), International Space Agency (ISA) and International Court of Air and Space Law(ICASL) in order to solute many legal problems in the field of the international air & outer space Law and Policy.

Acknowledgment

First of all, I have been researching the field of international aerospace law in depth for the past more 30 years, and so far, many of my research papers have been published in famous academic journals in Korea and abroad, and have made strong friendships with famous professors from all over the world.

It is indeed a great pleasure and honor for me to intimate and exchange our research results in the field of outer space law and policy with the worldwide famous Prof. Dr. Ram S. Jakhu (Canada), Prof. Dr. Fran G. von der Dunk (USA), Prof. Dr. Stephan Hobe and Dr. Marietta BenKö (Germany), Prof. Dr. William Maureen (Argentina), Prof. Dr. Irmgard Marboe (Austria), Prof. Dr. Li Shouping, Prof.Dr. Li Juqian, Secretary General Dr. Zhang Zhenjun, Prof. Dr. Gao Guozhu, Associate Prof. Dr. Maggie Qin Hauping, Associate Prof. Dr. Wang Guo Yu (China), Prof. Dr. Zhao Yun (Hong Kong), Prof. Kazuhiro Nakatani, Prof. Susumu Takai, Visiting Prof. Dr. Tomitaro Yoneda, Prof. Dr. Setsuko, Mr. Masahiko Sato, Dr. Yuri Takaya (Japan), Prof. Steven Freeland, Prof. Bin Li (Australia), Prof. Dr. V. S. Mani (India), Prof. Dr. Sun-Ihee Kim, Prof. Dr. Soon-Kil Hong, Prof. Dr. Kang-Bin Lee, Visiting Prof. Dr. Dong- Chun Shin, Prof. Dr. Young Jin Lee and Dr. Hong Je Cho (The Republic of Korea).

I would like to express my sincere appreciation to IGI Global Publisher as well as cooperation and effort of Ms. Jan Travers (Director of Intellectual Property & Contracts, IGI Global).

Chapter 1
UNCOPUOS and ILA National Space Legislation Based on the Space Treaties

ABSTRACT

This chapter begins with an introduction to the United Nations Committee on the Peaceful Use of Outer Space (UNCOPUOS) and the national space legislation based on the space treaties. This chapter describes the contents, research results, and perspective on the Model Law of National Space Legislation proposed by the UNCOPUOS and ILA. Especially, the author presents and comments on the contents of the legal text "Rules for Suborbital Activities Draft" proposed by the ILA Space Law Committee's Rapporteur. Outer space law is an area of the law that encompasses national and international law governing activities in outer space. Outer space law can be described as the body of law applicable to and governing space-related activities. The term "space law" is most often associated with the rules, principles and standards of international law appearing in the United Nation's five international space treaties.

1. 1. INTRODUCTION

Space has become an important sector in the world, which will develop with greater rapidity in the 21st century. Thanks to the expansion of the mobile communication, artificial intelligence and high technology industry, an increase in space activities of exploration and exploitation in the moon, Mars, Mercury, Venus, asteroid, Jupiter, Saturn and the Other Celestial Bodies, the implementation of the International Space Station (ISS) program, the space industry will grow continuously the market size of the globalization.

Outer space law is an area of the law that encompasses national and international law governing activities in outer space.[1] Outer space law can be described as the body of law applicable to and governing space-related activities. The term "space law" is most often associated with the rules, principles and standards of international law appearing in the United Nation's five international space treaties.

DOI: 10.4018/978-1-7998-7407-2.ch001

However, international space law also includes agreements, treaties, conventions, rules and regulations of international organizations (European Space Agency, International Telecommunications Union etc.), national laws, rules and regulations, executive and administrative orders, and judicial decisions. I would like to mention briefly the study for the model law of national space legislation proposed by the United Nations Committee on the Peaceful Use of Outer Space (UNCOPUOS) and International Law Association (ILA)[2] as the followings.

1.2. STUDY FOR THE MODEL LAW OF NATIONAL SPACE LEGISLATION PROPOSED BY UNCOPUOS AND INTERNATIONAL LAW ASSOCIATION

1.2.1. The Legal Committee and Legal Sub-Committee of UNCOPUOS[3]

The Legal Committee and Legal Subcommittee of the UNCOPUOS is the primary international forum for the development of laws and principles governing outer space.

The Legal Committee on the Peaceful Uses of Outer Space (COPUOS) was set up by the General Assembly in 1959 to govern the exploration and use of space for the benefit of all humanity for peace, security and development. The Legal Committee and Legal Subcommittee was instrumental in the creation of the five treaties and five principles of outer space.

1.2.2. Status and Application of the Five United Nations Treaties on the Outer Space

The Legal Committee has concluded five international legal instruments and five sets of major legal principles governing space-related activities.

These United Nations five treaties[4]provide for international legal principles such as the ① non-appropriation of outer space by any one country, ② the peaceful use of outer space, ③ the freedom of exploration, ④ liability for damage caused by space objects or space debris, ⑤ the safety and rescue of spacecraft and astronauts, ⑥ the prevention of harmful interference with space activities and the environment, ⑦ the notification and registration of space activities, ⑧ scientific investigation and exploitation of natural resources in outer space and ⑨ the peaceful settlement for the space disputes etc.

The legal principles adopted by the UN General Assembly address the application of international law and promotion of international cooperation and understanding in space activities, the dissemination and exchange of information through transnational direct television broadcasting via satellites and remote satellite observation of earth and general standards regulating the safe use of nuclear power sources necessary for the exploration and use of outer space.

The United States, the United Kingdom, Germany, Russia, Canada, China, Japan, India, Spain, Brazil, Australia, Mexico, Indonesia, the Republic of Korea etc. are a party to the four major international treaties related to space law: Space Treaty of 1967, Rescue Agreement of 1968, Liability Convention of 1972 and Registration Convention of 1975.

Pursuant to General Assembly resolution 64/86, agenda item 12, entitled "General exchange of information on national legislation relevant to the peaceful exploration and use of outer space," was considered in accordance with the multi-year work-plan for the period 2008-2011 adopted by the Committee on the Peaceful Uses of Outer Space at its fiftieth session.[5]

According to the Article VI of the 1967 Outer Space Treaty requests States to authorize and continuously supervise the activities of the space objects in their register, thus under their jurisdiction.[6] In view of the growing private space activities, national space legislation should cover a variety of problems. The Liability Convention of 192 covers only State liability and not the liability of a private actor.

Thus, only through national space legislation can it be assured that the responsible launching State may claim damage incurred by third-parties to the private space actor.

Therefore, questions of indemnification and insurance should be regulated through national space legislation. Thirty States in the world have adopted now the national space legislation as the following states;

Argentina, Australia, Austria, Belgium, Brazil, Canada, Chile, China, Denmark, Finland, France, Germany, India, Indonesia, Italy, Japan, Kazakhstan, Luxembourg, Norway, Philippine, Russian Federation, South Africa, Spain, Sweden, the Democratic Peoples Republic of Korea, The Netherlands, the Republic of Korea, Ukraine, the United Kingdom and the United States of America.

Domestically, as the space activities of both governmental and non-governmental entities in Korea are increasing, the Republic of Korea is developing a national space law system to implement these space law conventions. The Legal Subcommittee of the Committee on the Peaceful Uses of Outer Space held its fifty-eighth session at the United Nations Office at Vienna from 1 to 12 April 2019. [7]The session was conducted under the chairmanship of Andrzej Misztal (Poland). At its 976th meeting, on 1 April, 2019, the Legal Subcommittee adopted the following agenda:

5. Status and application of the five United Nations treaties on outer space.
7. National legislation relevant to the peaceful exploration and use of outer space.

The representatives from the 70 member States of the 58th Session's Legal Sub-committee was attended by the United States of America, the United Kingdom, Germany, France, Canada, China, Japan, Australia, India, Finland, Indonesia, the Republic of Korea etc.

Pursuant to UNCOPUOS General Assembly resolution 73/91, the Legal Subcommittee considered agenda item 5, entitled "Status and application of the five United Nations treaties on outer space", as a regular item on its agenda. At its 976th meeting, on 1 April, 2019, the Legal Subcommittee reconvened its Working Group on the Status and Application of the Five United Nations Treaties on Outer Space, with Bernhard Schmidt-Tedd (Germany) as Chair. At its 993rd meeting, on 11 April, 2019, the Legal Subcommittee endorsed the report of the Chair of the Working Group, contained in annex I to the present report.

The Legal Subcommittee noted that, as at 1 June 2019, the status of the five United Nations treaties on outer space was as follows:

1. The Outer Space Treaty had 109 States parties and had been signed by 23 additional States;
2. The Rescue Agreement had 98 States parties and had been signed by 23 additional States; three international intergovernmental organizations had declared their acceptance of the rights and obligations established under the Agreement;
3. The Convention on International Liability for Damage Caused by Space Objects (Liability Convention) had 96 States parties and had been signed by 19 additional States; four international intergovernmental organizations had declared their acceptance of the rights and obligations established under the Convention;

4. The Convention on Registration of Objects Launched into Outer Space (Registration Convention) had 69 States parties and had been signed by three additional States; four international intergovernmental organizations had declared their acceptance of the rights and obligations established under the Convention;
5. The Agreement Governing the Activities of States on the Moon and other Celestial Bodies (Moon Agreement) had 18 States parties and had been signed by eleven additional States.

Some delegations expressed the view that, as the five United Nations treaties on outer space formed the cornerstone of international space law, the UNCOPOUS Legal Subcommittee had a mandate to review its contents in the light of scientific and technical developments, and with a view towards addressing the current challenges presented by the diversification of space actors and the increasing privatization and commercialization of space activities.

The view was expressed that the five United Nations treaties on outer space continued to form the universal legal basis for present and future space exploration and use, and that the principles enshrined therein were equally valid for both countries with long-standing space programmes and emerging space actors.

The delegation expressing that view also expressed the view that the five United Nations treaties on outer space contributed to the safe and peaceful conduct of space activities and were for the benefit and in the interests of all countries.

The view was expressed that the likely reasons for the low adherence by States to the Moon Agreement were that the Agreement contained the pronouncement that the Moon and its natural resources were the "common heritage of humankind". The Moon Agreement proposed a regime to ensure that the Moon's natural resources were exploited for the benefit and in the interests of all countries.

The view was expressed that, although the status of the five United Nations treaties on outer space as cornerstones of international space law remained unchallenged, it had been clear to the drafters at the time of drafting that future scientific innovations and technological developments would necessitate improvements to the treaties.

The delegation expressing that view also expressed the view that, for the five United Nations treaties on outer space to remain relevant, the Subcommittee must consider the need for amendments and updates as well as promote even broader adherence to the legal regime governing outer space activities.[8]

1.2.3. National Legislation Relevant to the Peaceful Exploration and Use of Outer Space

The Legal Subcommittee of the UNCOPUOS held its forty-ninth session at the United Nations Office at Vienna from 22 March to 1 April 2010. At its 805th meeting, on 23 March 2010, the Legal Subcommittee of the UNCOPUOS reconvened its Working Group on National Legislation Relevant to the Peaceful Exploration and Use of Outer Space. The Working Group held six meetings, from 26 March to 1 April 2010.

The Working Group recalled that national regulatory frameworks represented different legal systems with either unified acts or a combination of national legal instruments and that States had adapted their national legal frameworks according to their specific needs and practical considerations. The Working Group continued its review of the following main issues for discussion:[9]

1. Reasons for States to enact national space legislation or the reasons for the absence of such legislation;
2. Scope of space activities targeted by national regulatory frameworks;
3. Scope of national jurisdiction over space activities;
4. Competence of national authorities in the authorization, registration and
5. supervision of space activities;
6. Conditions to be fulfilled for registration and authorization;
7. Regulations concerning liability;
8. Compliance and monitoring.

After a discussion of the proposal by the Chair on a tentative structure for the final report of the Working Group[10], the Working Group agreed that, upon completion of its multi-year work-plan, it should issue a comprehensive report on its work with the following structure:

1. Summary of the work conducted by the Working Group under its multiyear
2. work-plan
3. Overview of national space legislation
4. Findings of the Working Group
5. Conclusions[11]
6. I would like to introduce briefly the content of the Resolution adopted by the UN General Assembly on 11 December 2013 such as "recommendations on national legislation relevant to the peaceful exploration and use of outer space."

1. The scope of space activities targeted by national regulatory frameworks may include as appropriate, the launch of objects into and their return from outer space, the operation of a launch or re-entry site and the operation and control of space objects in orbit; other issues for consideration may include the design and manufacture of spacecraft, the application of space science and technology, and exploration activities and research;

2. The State, taking into account its obligations as a launching State and as a State responsible for national activities in outer space under the United Nations treaties on outer space, should ascertain national jurisdiction over space activities carried out from territory under its jurisdiction and/or control; likewise, it should issue authorizations for and ensure supervision over space activities carried out elsewhere by its citizens and/or legal persons established, registered or seated in territory under its jurisdiction and/ or control, provided, however, that if another State is exercising jurisdiction with respect to such activities, the State should consider forbearing from duplicative requirements and avoid unnecessary burdens.

Pursuant to UN General Assembly resolution 73/91, the Subcommittee considered agenda item 7, entitled *"National legislation relevant to the peaceful exploration and use of outer space"*, as a regular item on its agenda on 1~12 April 2019. The representatives of Armenia, Australia, Chile, Finland, India, Indonesia, Japan, Mexico, the United Arab Emirates, the United Kingdom and the United States made statements under agenda item 7. During the general exchange of views, statements relating to the item were made by the representatives of other member States.

The UNCOPUOS Legal Subcommittee reiterated that it was important to take into account the emerging trend of non-governmental entities engaging in outer space activities. In that connection, States needed to ensure, through their national legal frameworks, that the outer space activities of those entities were in compliance with the United Nations treaties on outer space, in order to ensure the safety and security of those activities.

The Legal Subcommittee noted that national space policies and implementing regulations were increasingly addressing issues raised by the rising number of non-governmental entities conducting space activities.

The Legal Subcommittee also noted that those activities were aimed at improving the management and regulation of space activities; reorganizing national space agencies; increasing the competitiveness of governmental and non-governmental organizations in their space activities; increasing the involvement of academia in policy formulation; improving responses to challenges posed by the development of space activities, in particular those relating to the management of the space environment; and improving compliance with international obligations.

The view was expressed that, when drafting national space law, international regulatory frameworks, including the United Nations treaties and principles on outer space, the ITU Constitution and Convention and the ITU Radio Regulations, and certain non-binding instruments, including the Space Debris Mitigation Guidelines of the Committee on the Peaceful Uses of Outer Space, should be taken into consideration in order to guarantee the safety and sustainability of outer space activities.

The view was expressed that as countries consider their national space policies, including their legal and regulatory frameworks, all delegations should agree to respect the history of humans on the Moon, including the significance of Luna 2 in 1959, the first human landing on the Moon in 1969, the first soft landing on the far side of the Moon in 2019 and multiple other missions.

The delegation expressing that view also expressed the view that all States should acknowledge the significance that these historic activities have had for society, and called on States to recognize and respect the historic, cultural and scientific importance of these missions and the traces they have left on the Moon as they develop their national policies.

The Legal Subcommittee agreed that it was important to continue to regularly exchange information on developments in the area of national space-related regulatory frameworks. In that regard, the Legal Subcommittee encouraged member States to continue to submit to the Secretariat texts of their national space laws and regulations and to provide updates and input for the schematic overview of national regulatory frameworks for space activities.[12] The Legal Subcommittee under the UNCOPUOS was scheduled to hold its 59th session at Vienna from 23 March to 3 April 2020. This session as scheduled was cancelled by the COVID-19 epidemic.[13]

1.3. ACTIVITY OF SPACE LAW COMMITTEE UNDER THE ILA[14]

The obligation of authorization and continuous supervision of space activities by private actors should also be regulated for the sole reason that it is the only way for governments to control these activities. Since the 2004 Conference in Berlin, the International Law Association(ILA), Space Law Committee has been dealing with the national space legislation.

Furthermore, at the 2008 ILA Conference in Rio de Janeiro the Rapporteur had submitted a report on national space legislation, which is currently very active in the exchange of opinions on national space legislation.

As the Space Law Committee of ILA at Toronto and Rio de Janeiro, its work is directed towards the formulation and final adoption of a model for national space legislation, depending on the progress of its work in 2010 at its Hague Conference and/or in 2012 in Sofia, Bulgaria[15].

Such a document might still play a positive role during the final stage of discussions on the same topic at the Legal Subcommittee of UNCOPUOS. During the ILA 74th Conference, last August 2010 at The Hague, the Space Law Committee discussed, inter alia, a "*Draft Model Law on the National Space Legislation* (NSL)" presented by the Committee Rapporteur Professor Stephan Hobe (Germany). This Draft Model Law on NSL is composed of nine Articles as the following;

1. Requirements for Private Space Activities,
2. Contents of the License,
3. Transfer of the License,
4. Insurance,
5. Registration,
6. Supervision,
7. Environmental Assessment,
8. Compensation,
9. Dispute Settlement.

1.3.1. Perspective of the ILA Model Law on the National Space Legislation

The International Law Association (ILA) was founded in Brussels in 1873. Its objectives, under its Constitution, are "the study, clarification and development of international law, both public and private, and the furtherance of international understanding and respect for international law". The ILA has consultative status, as an international non-governmental organisation, with a number of the United Nations specialised agencies.

The headquarter of ILA is located now in London, the U.K. and ILA has 63 countries' branch including U.S.A. the U.K. France, Germany, Russia, Canada, Japan, India, Spain, Mexico, Brazil, Argentina, Australia, Republic of Korea etc. ILA members is now 4,503 professors, scholars, Judges, lawyers and government officials etc. and IlA has 20 Active Committee including Space Law Committee. I am now a member of the ILA Space Law Committee since 1994.[16]

The ILA Space Law Committee was set up in New York in 1958 and has worked and met without interruption to date. Its present officers of ILA Space Law Committee are Chair, Professor Stephan Hobe (Germany) and Co-chair, Professor Maureen Williams (Headquarters). The Committee is honored to be, since 1990, a permanent observer to the Committee on the Peaceful Uses of Outer Space (COPUOS) and both its Subcommittees, to which it reports annually.

I received a copy of the "*Proposed Model Law on National Space Legislation (NSL)*" from Chair of the Space Law Committee of ILA Prof. Maureen Williams by E-mail on November 1, 2010. This Proposed Model Law on NSL also is composed of thirteen Articles as the following;

① Scope of the law, ② Definitions - Use of Terms, ③ Authorization and supervision, ④ Conditions for authorization, ⑤ Transfer of License, ⑥ Insurance, ⑦ Registration, ⑧ Supervision, ⑨ Environmental Assessment, ⑩ Mitigation of Space Debris, ⑪ Liability and Compensation, ⑫ Dispute Settlement, ⑬ Sanctions.

It is indeed a great necessary and desirable for us to regulate a standard and enact a space national legislation in order to unify the different legal systems among the 192 countries in the global community. An ILA Draft for the Space Model Law should be finalized and submitted for endorsement by the ILA Bulgaria Conference in 2012.

The Sofia Guidelines for a Model Law on National Space Legislation were adopted by the 75th ILA Conference on 30 August 2012 as resolution June, 2012.

Our latest activity, to which we have been particularly dedicated together with the Committee members, has focused on our current terms of reference, ranging from 2014 to 2020 when our Final Report should be submitted to the 2020 ILA Kyoto (Japan) Conference.

Our latest ILA World Conference was held in August 2018 in Sydney where, at its Working Session, the Space Law Committee presented and discussed a number of central topics included in the on-going mandate and a couple of specific issues further developed in the course of the mandate (Part 1 of the Sydney Report, by the Chair) and a first reading of the Draft Model Guidelines dealing with Suborbital Activities (Part 2 of the Sydney Report prepared by the General Rapporteur).

Given that these meetings were held less than a year ago it was thought opportune to highlight these –at times- controversial new issues at the forthcoming meeting of the LSC in Vienna in April 2019. This would be the last stage of the mandate of the ILA Space Law Committee, on the threshold of the Kyoto Conference where our Final Report, as announced earlier, is to be submitted.

1.3.2. The Current Terms of Reference of the ILA Space Law Committee: 2014- 2020

The mandate of the Space Law Committee for 2014-2016 (and later extended by the ILA Council for another four years, namely until 2020), reads as follows. [17]

The Committee has an established and continuing relationship with the UN institutions dealing with space law and, in addition to its on-going work with these bodies, the topics under the new mandate (2012-2016) include:

1. Dispute settlement and the 2011 PCA Rules on Arbitration,
2. Legal aspects of suborbital flight,
3. Satellite data in international litigation and
4. New developments on space debris.

Moreover, the Committee will keep a general watching brief over further developments during the four-year mandate.

The First Space Law Committee Report under this new mandate was submitted in 2014 to the Seventy-Sixth Conference in Washington, DC. It was followed by a Second Report in 2016 (Johannesburg Conference), a Third and latest Report in 2018 (Sydney Conference) and a Fourth and Final one to be submitted in August 2020 (Kyoto Conference).

1.3.3. The ILA 79th 2020 Kyoto, Japan International Law Conference

The ILA 79th International Law Conference will take place from 23-27 August, 2020 in Kyoto, Japan. The 2020 Kyoto ILA Space Law Committee will be discussed the contents of *"Rules for suborbital activities Draft"* proposed by Prof. Dr. Stephan Hobe,[18] Rapporteur of the ILA Space Law Committee.[19]

The ILA 79th Kyoto International Law Conference was postponed by "the Corona Virus 19 pandemic ".

1.3.4. Chronology of the Work of the ILA Space Law Committee on the Topic of Suborbital Flight Activities

For the work of the ILA Space Law Committee, in 2013, the Rapporteur was tasked to draft a report on *"Legal problems of private commercial manned suborbital flights"*.

In 2016, the Rapporteur prepared a second report for the 77th biennial conference of the ILA in Johannesburg (South Africa) which elaborated on the points laid down in the 2014 report and provided an update of more recent developments concerning suborbital activities, such as definitional issues, the role of existing technological concepts and projects for determining applicable rules, state practice, liability aspects and the avenues for the possible evolution of commercial suborbital touristic flights to a transportation market.

During the 77th biennial conference in Johannesburg, South Africa, in 2016 the Rapporteur has been mandated to formulate draft rules for suborbital activities which were presented and discussed during the 78th biennial conference in Sydney, Australia, in 2018.[20]

1.3.5. General Overview of the Legal Problems Related to Suborbital Activities

The regulation of suborbital activities poses a long list of unsolved legal issues which have been discussed in for a such as International Civil Aviatiopn Organization (ICAO), UNCOPUOS and United Nations Office for Outer Space Affairs (UNOOSA) on the international level, and by national authorities, e.g. the U.S. Federal Aviation Administration.

They include:

- the lack of legally binding definitions clarifying relevant terms such as, e.g.,'suborbital vehicle', 'suborbital flight' and 'suborbital activity';
- the un-clearity regarding the applicable international law;
- the relevance of air law and space law;
- the scope and possible elements of a potential sui generis legal order;
- the question of a competent authority for the regulation;
- the relationship between national legislation and the applicable international law to suborbital activities;
- the scope of a comprehensive liability and insurance regime for Suborbital activities;
- the need for a registration regime for suborbital activities;
- the need for an extensive authorization regime guaranteeing the "air/ spaceworthiness" of suborbital vehicles as well as the coordination between air traffic and space traffic, incl. objects using the so-called meso-space e.g. 4HAPS);
- policies applicable for passengers onboard of a suborbital vehicle;

- the status of different categories of persons participating in suborbital activities (e.g. pilots, trained crew, passengers);
- various aspects of space transportation as an application of suborbital activities

1.3.6. Legal text of "Rules for Suborbital Activities Draft" Proposed by ILA Space Law Committee's Rapporteur

Article 1, Definitions

1. suborbital activity is an activity aimed at transportation ofpersons or cargo into
2. outer space involving a parabolic flight not sending an object into Earth orbit.
3. A suborbital vehicle is a vehicle used for a suborbital activity.
4. A suborbital operator is the entity, private or governmenta performing a suborbital activity under the jurisdiction of one State.

Article 1, para. a), Definition of 'Suborbital Activity'

A suborbital activity is an activity aimed at transportation of persons or cargo outer space involving a parabolic flight not sending an object into Earth orbit.

Article 1, para. b) Definition of 'Suborbital Vehicle'

A suborbital vehicle is a vehicle used for a suborbital activity.

Article 1, para. c) Definition of 'Suborbital Operator'

A suborbital operator is the entity, private or governmental, performing a suborbital activity under the jurisdiction of one State.

Article 2, Applicability of International Law

International law is applicable to suborbital activities.

Article 3, Authorizing State

Every suborbital activity must be authorized by the State under whose territorial jurisdiction the activity is started.

Article 4, Common Elements of National Authorization Regimes

The criteria which States shall apply when deciding on an authorization of suborbital vehicles include:

1. Main purpose of the activity;
2. Space and airworthiness of the person;

3. Route description: altitude /trajectory/ time of flight/start and landing location;
4. Obligation for coordination with ground air traffic control;
5. Type of cargo / number of crew members and flight participants;
6. Safety requirements for the vehicle, air/space worthiness, disclosure of main Design features for the vehicle;
7. Qualification of crew members;
8. Registration of the suborbital in a national register;
9. Requirement of an insurance of the vehicle and of any person participating in the activity through the operator;
10. Conditions for recourse between the state and the suborbital operator.

Article 5, Registration of the Suborbital Activity and the Suborbital Vehicle

1. A suborbital activity and the suborbital vehicle must be registered in a national registry established and maintained by the authorizing State.
2. The suborbital operator must provide the following details for the registration.
    a. Location of the launch/take-off;
    b. Date of launch;
    c. Type and registration number of the vehicle;
    d. Apogee, perigee and trajectory;
    e. Location of landing
    f. Date of landing;
3. The information provided for in the national registry must be furnished as soon as possible to the Secretary General of the United Nations who maintains an international registry.

Article 6, Liability and Insurance

1. The authorizing State is liable for any damage caused by the suborbital activity to persons or property.
2. The term 'damage' means loss of life, personal injury or the impairment
3. impairment of health; or loss of or damage to property of persons, natural or judicial, or damage to property of international intergovernmental organisations; or environmental damage.
4. The authorising State can have recourse against the insurer of the suborbital operator.

Article 7, Crew and Passengers in a Suborbital Activity

1. Definition of crew member of the crew is any trained employee of a suborbital operator, or of a contractor or suborbital subcontractor of a suborbital perator, who performs activities in the course of that employmen directly relating to the launch, re-entry, or other operation of or in a suborbital vehicle that carries human beings.
2. Definition of passengers is any person on board a suborbital vehicle during the duration of the suborbital activity who is not a member of the crew.
3. Definition of cargo is any freight that is on board for the purpose of being transported.

Article 8, Suborbital Facilities

Details concerning ground facilities used for suborbital activities shall be regulated in specific statutory bylaws.

Prof. Dr. Stephan Hobe (Chair of Space Law Committee, ILA) suggest November 2021 for this ILA Space Law meeting, with a view to having enough momentum for the adoption of our *Draft Rules for suborbital (non-orbital) spaceflight activities* so that they can be presented to the Executive Committee at the next biennial Conference in Lisbon, Portugal in 2022 for adoption.

1.3.7. Comment on the "Rules for Suborbital Activities Draft"

I think that this "*Rules for Suborbital Activities Draft* "has a meaningful and excellent Draft. This "*Rules for Suborbital Activities Draft* "was commented by famous space law professor such as Prof. Irmgard Marboe (Austria), Mr. Niklas Hedman (UNOOSA), Prof. Dr. Frans von der Dunk (USA) and Prof. Steven Freeland (Australia).

I would like to comment this "Rules for suborbital activities Draft" as the following;

Article 6, Liability and Insurance

1. The authorizing State or Suborbital Flight Operator is liable for any damage caused by the suborbital activity to person or property.

Comment

In the near future, suborbital flight tour will be accomplished and increased by suborbital flight vehicle, because it is not a space flight. American private companies were hard at work trying to make space tourism a profitable proposition by developing suborbital vehicles designed to take passengers to an altitude of 100 km (62 miles). XCOR's Lynx(USA) suborbital space plane. The Lynx is XCOR's entry into the commercial reusable launch vehicle (RLV) market.

This two-seat, piloted space transport vehicle will take humans and payloads on a half-hour suborbital flight to 100 km (330,000 feet) and then return safely to a landing at the takeoff runway. Therefore, when suborbital flight operator (private company etc.) has negligence in launching process by suborbital light vehicle, mis-manufacture of suborbital flight vehicle or damaged by a third party due to a collision between space debris and suborbital flight vehicle, the suborbital flight operator (private company etc.) must indemnify compensation for damage to the victims.

In this reason I hope that drafter would insert a phrase such as suborbital flight operator (private company etc.) into the text of Article 6 as the abovementioned comment.

Article 6, Liability and Insurance

Suborbital flight operator must cover a personal, property and liability insurance before launching the suborbital flight vehicle.

Comment

I wish that drafter would add 4 paragraph sentence into Article 6 as follows.

When a suborbital flight operator launches a suborbital flight vehicle or in case of accidents such as a sudden drop, collision, etc., damages to personal, property, or third parties must be compensated. In this reason suborbital flight operator must cover certainly the compulsory insurance before launching a suborbital flight vehicle.

ADDITIONAL READING

Cheng, C.-J. (1993). *The Use of Airspace and Outer Space for all Mankind in the 21st Century*. Kluwer Law International.

Lele. (2012). *Decoding the International Code of Conduct for Outer Space Activities*. Pentagon Security International.

KEY TERMS AND DEFINITIONS

IATA: The International Air Transport Association (IATA) supports aviation with global standards for airline safety, security, efficiency, and sustainability. Consisting in 2016 of 290 airlines, primarily major carriers, representing 117 countries, the IATA's member airlines account for carrying approximately 82% of total available seat miles air traffic. It is headquartered in the city of Montréal, Canada, with Executive Offices in Geneva, Switzerland.

ICAO: The International Civil Aviation Organization (ICAO) is a specialized agency of the United Nations. It changes the principles and techniques of international air navigation and fosters the planning and development of international air transport to ensure safe and orderly growth. Its headquarters is located in the *Quartier International* of Montreal, Quebec, Canada. The ICAO Council adopts standards and recommended practices concerning air navigation, its infrastructure, flight inspection, prevention of unlawful interference, and facilitation of border-crossing procedures for international civil aviation. ICAO defines the protocols for air accident investigation that are followed by transport safety authorities in countries signatory to the Chicago Convention on International Civil Aviation.

ILA: The International Law Association (ILA) is a non-profit organization based in Great Britain that - according to its constitution - promotes "the study, clarification and development of international law" and "the furtherance of international understanding and respect for international law". The ILA was founded in Brussels in 1873 and its present-day headquarters are in London, UK. Currently, the ILA has 20 active committees and 8 study groups that analyze specific facets of private and public international law. I am now a member of Space Law Committee of ILA. There are over 4,500 active ILA members around the world.

UNCOPUOS: The United Nations Committee on the Peaceful Uses of Outer Space (COPUOS) was established in 1958 as an *ad hoc* committee of the United Nations (UN). In 1959, it was formally established as a permanent body by United Nations resolution 1472 (XIV). The mission of COPUOS is "to review the scope of international cooperation in peaceful uses of outer space. As of 2019, UNCOPUOS member states is 92 countries and it's head quarter is located in Vienna, Austria.

ENDNOTES

1 https://en.wikipedia.org/wiki/Space_law
2 https://www.ila-hq.org/
3 http://www.oosa.unvienna.org/oosa/COPUOS/copuos.html
4 Space Treaty of 1967, the Rescue Agreement of 1968, the Liability Convention of 1972, the Registration on Convention of 1975 and Moon Agreement of 1979.
5 http://www.oosa.unvienna.org/pdf/reports/ac105/AC105_942E.pdf
6 S. Hobe/B. Schmidt-Tedd/K.-U. Schrogl (eds.), *Cologne Commentary on Space Law*, vol. 1, Cologne, . 2009, pp. 103.
7 https://www.unoosa.org/oosa/events/data/2019/58th_session_of_the_legal_subcommittee.html
8 https://www.unoosa.org/oosa/events/data/2019/58th_session_of_the_legal_subcommittee.html
9 see A/AC.105/935, annex III, paras. 7 and 18
10 A/AC.105/C.2/2010/CRP.16
11 http://www.oosa.unvienna.org/pdf/reports/ac105/AC105_942E.pdf
12 https://www.unoosa.org/oosa/events/data/2019/58th_session_of_the_legal_subcommittee.html
13 https://www.unoosa.org/oosa/en/ourwork/copuos/lsc/2020/index.html
14 https://www.ila-hq.org/en/committees/index.cfm/cid/29
15 https://www.unil.ch/dip/page82217_en.html
16 https://www.ila-hq.org/index.php/committees
17 https://www.ila-hq.org/index.php/committees
18 Director, Institute of Air Law, Space Law and Cyber Law, Managing Co-Director International Investment Law Centre Cologne, University of Cologne, Germany.
19 Ms.Rada Popova, senior research fellow at the Institute of Air Law, Space Law and Cyber Law at the University to Cologne, for her involvement in the preparation of this reportand to the Committee members for their input.
20 Report from the working session of the Space Law Committee, 78th biennial Conference of the International Law Association, 19-24 August 2018, Sydney, Australia.

Chapter 2
Main Contents and Comment on the 1967 Outer Space Treaty

ABSTRACT

This chapter introduces the ratification by member states and main contents of the 1967 Outer Space Treaty (Title: Treaty on Principles Governing the Activities of States in the Exploration and Use of Outer Space, Including the Moon and Other Celestial Bodies). Furthermore, the author explains the reason it the contents of the 1967 Outer Space Treaty should be amended. The treaty was opened for signature in the United States, the United Kingdom, and the Soviet Union on 27 January 1967, and entered into force on 10 October 1967. As of June 2020, 110 countries are parties to the treaty, while another 23 have signed the treaty but have not completed ratification.

2.1. INTRODUCTION

First of all I would like to introduce briefly the main contents and comment of the United Nations 1967 Outer Space Treaty as well as UN four space treaties as the followings. The UN General Assembly (UNGA) took a particular interest in space affairs. It adopted resolutions on space issues, which ultimately culminated in the five UN Treaties on outer space.

Apart from these major multilateral Treaties, the UNGA also adopted various other special conventions relating to space-based activities, including the 1963 Treaty Banning Nuclear Weapon Tests in the Atmosphere, in Outer Space and Under Water, the 1977 Convention on the Prohibition of Military and Any Other Hostile Use of Environmental Modification Techniques, and the Convention and Regulation of the International Telecommunications Union.

The UN General Assembly adopted Resolution 1962 (XVIII) on 13 December 1963, the Declaration of Legal Principles Governing the Activities of States in the Exploration and Use of Outer Space. In June 1966 the United States and the Soviet Union submitted draft treaties on the uses of space to the United Nations. These were reconciled during several months of negotiation in the Legal Subcommittee of the

DOI: 10.4018/978-1-7998-7407-2.ch002

UN Committee on the Peaceful Uses of Outer Space, and the resulting document was endorsed by the UN General Assembly on Dec. 19, 1966. The Declaration set out various fundamental principles which were expanded upon in the Outer Space Treaty of 1967.

2.2. RATIFICATION OF OUTER SPACE TREATY

Outer Space Treaty (OST), formally *"Treaty on Principles Governing the Activities of States in the Exploration and Use of Outer Space, Including the Moon and Other Celestial Bodies"* of 1967, is a treaty that forms the basis of international space law. The Outer Space Treaty was opened for signature in the United States, the United Kingdom, and the Soviet Union on 27 January 1967, and entered into force on 10 October 1967.

Figure 1.
Source: https://en.wikipedia.org/wiki/Outer_Space_Treaty

As of June 2020, 110 States including the USA, the United Kingdom, Canada, France, Germany, Russia, China, Japan, Italy, Spain, Brazil, Argentina, Mexico, Australia, India, Indonesia, the Republic of Korea and North Korea etc. are parties to the treaty, while another 23 States have signed the Treaty but have not completed ratification.

In addition, the Republic of China (Taiwan), which is currently only recognized by 14 UN member states, ratified the treaty prior to the United Nations General Assembly's vote to transfer China's seat to the People's Republic of China (PRC) in 1971.[1]

2.3. MAIN CONTENTS OF OUTER SPACE TREATY

The OST represents the basic legal framework of international space law. Among its principles, it bans States Parties to the Space Treaty from placing nuclear weapons or any other weapons of mass destruction in orbit of Earth, installing them on the moon or any other celestial body, or to otherwise station them in outer space.

It exclusively limits the use of the moon and other celestial bodies to peaceful purposes and expressly prohibits their use for testing weapons of any kind, conducting military maneuvers, or establishing military bases, installations, and fortifications (Art. IV). However, the Space Treaty does not prohibit the placement of conventional weapons in orbit. The Space Treaty also states that the exploration of outer space shall be done to benefit all countries and shall be free for exploration and use by all the States.

The OST also stresses the promotion of international cooperation in the exploration and use of space, the moon and other celestial bodies. Furthermore, the OST establishes the principles of non-appropriation and the fact that no state can claim sovereignty of or occupy outer space, the moon or any other celestial body.

Moreover, the OST prescribes the role of non-governmental activities in outer space, the moon and other celestial bodies, namely that their activities continue under the supervision of the appropriate state party to the Space Treaty.

The Space Treaty explicitly forbids any government from claiming a celestial resource such as the moon or a planet, claiming that they are *"the common heritage of mankind."* [2] Article II of the OST states that "*Outer space, including the moon and other celestial bodies, is not subject to national appropriation by claim of sovereignty, by means of use or occupation, or by any other means*". However, the State that launches a space object retains jurisdiction and control over that object.[3]The State is also liable for damages caused by their space object and must avoid contaminating space and celestial bodies.[4]Seeking clearer guidelines, private companies in the US prompted the US government and legalized space mining in 2015 by introducing the US Commercial Space Launch Competitiveness Act of 2015.[5]

Similar national legislations legalizing extraterrestrial appropriation of resources are now being replicated by other nations, including Luxembourg etc. This has created a controversy on legal claims for mining for profit.

2.4. RESPONSIBILITY FOR ACTIVITIES IN SPACE

Article VI of the OST deals with international responsibility, stating that "*the activities of non-governmental entities in outer space, including the moon and other celestial bodies, shall require authorization and continuing supervision by the appropriate State Party to the Treaty*" and that States Parties shall bear international responsibility for national space activities whether carried out by governmental or non-governmental entities.

The responsibility for compliance of the activities of an inter-governmental organization is borne by the relevant organization and states party to the Space Treaty participating in such an organization. Finally, the OST deals with liability, the position of astronauts, as well as states' responsibility to inform the UN Secretary General and the international scientific community of the nature, conduct, locations and results of their activities in outer space.[6]

2.5. COMMENT ON THE OUTER SPACE TREATY

The Space Treaty has not solved the problem of the meaning and interpretation to be given to the term "peaceful." In determining the future uses of the moon and other celestial bodies, the present analysis suggests that if we desire to introduce some measure of certainty it would be better to abandon such a general and undefined phrase and instead identify permissible or prohibited activities. The relevant Treaty provisions are indicative of a partial attempt to move in this direction.

However, it seems somewhat unfortunate that they do not provide for further identification of specific problem areas so as to avoid endless arguments as to whether the word "peaceful" means nonmilitary or nonaggressive or perhaps something else.

Even though questions of interpretation would still remain concerning the scope of prohibited activities, these could be resolved by a definition of important terms in a future outer space convention.[7]

The space law system, indeed, does not provide any specific rule, relating to the exploitation of extraterrestrial resources, which have been generally accepted by States.

According to the 1967 Outer Space Treaty and the 1979 Moon Agreement, these two instruments does not offer an adequate legal framework which is able to ensure the safe, orderly and peaceful development and extraction from the natural resources in the Moon and other celestial bodies.

On one side, the Outer Space Treaty does not contain any mention on the essence and duration of mining right in the exploitation space natural resources in moon and other celestial bodies.

The space law system, indeed, does not provide any specific rule, relating to the exploitation of extraterrestrial resources, which have been generally accepted by many States.

According to the "1967 Outer Space Treaty and 1979 Moon Agreement" these two instruments does not offer an adequate legal framework which is able to ensure the safe, orderly and peaceful development and extraction from the natural resources in the Moon and other celestial bodies.

On one side, the Outer Space Treaty does not contain any mention on the essence and duration of mineral right in the exploitation and development of the space natural resources in moon, Mars, asteroids, Jupiter, Saturn and other celestial bodies.

It is a great necessary and desirable for us to revise the 1967 Outer Space Treaty in order to regulate the meaning, license and duration of mining right for exploring and developing of the natural resources in the Moon, Mars, asteroids and Other Celestial Bodies.

ADDITIONAL READING

Christol, C. Q. (1991). *Space Law: Past, Present and Future*. Kluwer.

Jasani, B. (1991). *Outer Space, A source of Conflict or Co-operation?* United Nations University Press.

KEY TERMS AND DEFINITIONS

International Telecommunications Union: The International Telecommunication Union (ITU), is a specialized agency of the United Nations responsible for all matters related to information and communication technologies. Established in 1865 as the International Telegraph Union, it is one of the oldest international organizations in operation. Based in Geneva, Switzerland, the ITU's global membership includes 193 countries and around 900 business, academic institutions, and international and regional organizations.

UNOOSA: The United Nations Office for Outer Space Affairs (UNOOSA) is tasked with promoting the peaceful use and exploration of space through international cooperation. Part of the United Nations

Secretariat, it works with any of the 193 UN Member States to establish or strengthen the legal and regulatory frameworks for space activities, and assists developing countries in using space science and technology for sustainable socioeconomic development. The Office was created in 1958 to assist and advise the *ad hoc* Committee on Peaceful Uses of Outer Space (COPUOS), which was established by the UN General Assembly to discuss the scientific and legal aspects of exploring and using outer space to the benefit of humankind. The Committee became permanent the following year, with UNOOSA undergoing several structural changes before its relocation in 1993 to the United Nations Office in Vienna, Austria.

US Commercial Space Launch Competitiveness Act: The United States Government updated US commercial space legislation in 2015 with the passage of the Commercial Space Launch Competitiveness Act, sometimes referred to at the *Spurring Private Aerospace Competitiveness and Entrepreneurship (SPACE) Act of 2015*. The update to US law explicitly allows US citizens to "engage in the commercial exploration and exploitation of 'space resources' [including ... water and minerals]." The right does not extend to biological life, so anything that is alive may not be exploited commercially. The Act further asserts that "the United States does not [(by this Act)] assert sovereignty, or sovereign or exclusive rights or jurisdiction over, or the ownership of, any celestial body."

ENDNOTES

1. https://en.wikipedia.org/wiki/Outer_Space_Treaty
2. Jennifer Frakes, The Common Heritage of Mankind Principle and the Deep Seabed, Outer Space, and Antarctica: Will Developed and Developing Nations Reach a Compromise? Wisconsin International Law Journal (2003, 21 at 409).
3. Article 8 of the 1967 Outer Space Treaty.
4. Article7 of the 1967 Outer Space Treaty.
5. *"U.S. Commercial Space Launch Competitiveness Act"*. Act No. H.R.2262 of 5 December 2015. 114th Congress (2015-2016) Sponsor: Rep. McCarthy, Kevin.
6. https://thespacereview.com/article/3155/1
7. http://digitalcommons.law.uga.edu/cgi/viewcontent.cgi?article=2191&context=gjicl

Chapter 3
Main Contents and Comment on the 1968 Rescue Agreement

ABSTRACT

This chapter describes the historical background, basic provisions, main contents, and key changes of the 1968 Space Rescue Agreement (Title: Agreement on the Rescue of Astronauts, the Return of Astronauts and the Return of Objects Launched Into Outer Space). The author points out the legal problems and the solution on the 1965 Rescue Agreement. The Rescue Agreement requires that any state party that becomes aware that the personnel of a spacecraft are in distress must notify the launching authority and the Secretary General of the United Nations. The UN General Assembly adopted the text of the Rescue Agreement on 19 December 1967 through Resolution 2345 (XXII). The Agreement opened for signature on 22 April 1968, and it entered into force on 3 December 1968. As of January 2019, 98 states have ratified the Rescue Agreement, 23 have signed, and three international intergovernmental organizations (the European Space Agency, the Intersputnik International Organization of Space Communications, and the European Organisation for the Exploitation of Meteorological Satellites) have declared their acceptance of the rights and obligations conferred by the agreement.

3.1. INTRODUCTION

The 1968 Rescue Agreement *(Title: The Agreement on the Rescue of Astronauts, the Return of Astronauts and the Return of Objects Launched into Outer Space),* also referred to as an international agreement setting forth rights and obligations of states concerning the rescue of persons in space. This Rescue Agreement was created by a 19 December 1967 consensus vote in the United Nations General Assembly (Resolution 2345 (XXII)). It came into force on 3 December 1968.

Its provisions elaborate on the rescue provisions in Article V of the 1967 Outer Space Treaty. Despite containing more specificity and detail than the rescue provision in Article V of the Outer Space Treaty, the Rescue Agreement still suffers from vague drafting and the possibility of differing interpretation. By the end of 1967, space diplomacy proved capable of accomplishing an international agreement at the speed of space travel itself.

DOI: 10.4018/978-1-7998-7407-2.ch003

On the very last day before the end of its 22nd Session, the United Nations General Assembly was confronted with an Agreement on the Rescue and Return of Astronauts and the Return of Objects Launched into Outer Space and almost simultaneously adopted a resolution in which it "commends" the Agreement.

United Nations Committee had not been able to produce in five years of hard work, emerged suddenly from private negotiations between the United States and the Soviet Union. Although many of the Agreement's particular provisions submitted by the two space powers had their origin in earlier texts discussed in the UN Outer Space Committee and its legal subcommittee, the fact remains that it was a bilateral draft which was rushed through the multilateral arena.[1]

3.2. CONCLUSION OF THE RESCUE AGREEMENT

The UN General Assembly adopted the text of the Rescue Agreement on 19 December 1967 through Resolution 2345 (XXII). The Agreement opened for signature on 22 April 1968, and it entered into force on 3 December 1968. As of January 2019, 98 States including the USA, the United Kingdom, Canada, France, Germany, Russia, China, Japan, Italy, Spain, Brazil, Argentina, Mexico, Australia, India, Indonesia, the Republic of Korea etc.

Have ratified the Rescue Agreement, 23 have signed, and three international intergovernmental organizations (the European Space Agency, the Intersputnik International Organization of Space Communications, and the European Organisation for the Exploitation of Meteorological Satellites) have declared their acceptance of the rights and obligations conferred by the agreement.[2]

3.3. BASIC PROVISIONS ON THE RESCUE AGREEMENT

The Rescue Agreement requires that any state party that becomes aware that the personnel of a spacecraft are in distress must notify the launching authority and the Secretary General of the United Nations. The Rescue Agreement essentially provides that any state that is a party to the agreement must provide all possible assistance to rescue the personnel of a spacecraft who have landed within that state's territory, whether because of an accident, distress, emergency, or unintended landing.

If the distress occurs in an area that is beyond the territory of any nation, then any state party that is in a position to do so shall, if necessary, extend assistance in the search and rescue operation.

3.4. KEY CHANGES ON THE RESCUE AGREEMENT

3.4.1. Parties Entitled to Be Rescued

The Outer Space Treaty of 1967 states simply that astronauts are to be rendered all possible assistance by state parties to the treaty. The Outer Space Treaty does not provide a definition for the term "astronaut", and as a result it is unclear whether this provision applies to, for example, a space tourist-a person who clearly has not received the training of a traditional astronaut. The Rescue Agreement adds some clarity to the issue by referring to the "personnel of a spacecraft" rather than "astronauts". However, this phrase again leaves uncertain whether someone simply along for the ride—such as a tourist on a Virgin Galactic flight—would be considered part of the "personnel of a spacecraft".

3.4.2. Compensation for Recovery of a Space Object

In the event that a space object or its parts land in the territory of another state party, the state where the object lands is required (upon the request of the launching authority) to recover the space object and return it to the launching authority. The Rescue Agreement provides that the launching state must then compensate the state for the costs incurred in recovering and returning the space object.

3.4.3. Rescue in Space

For many years, the idea of a rescue while actually in space was viewed as somewhat unrealistic, since it was unlikely that any state would have a craft ready to reach astronauts in distress in time to rescue them. The situation has evolved over time, however. Mir and later the International Space Station have each maintained docked Russian Soyuz spacecraft to be used as an escape mechanism in the event of an on-orbit emergency.

A significant shift in attitudes toward in-flight rescues came as a result of the Space Shuttle Columbia disaster, after which NASA took steps to prepare the STS-3xx or *Launch on Need* missions to provide for rescue in certain scenarios. However, no rescue mission was subsequently required through the remaining duration of the Space Shuttle program.

3.5. COMMENT

The Rescue Agreement has been criticized for being vague, especially regarding the definition of who is entitled to be rescued and the definition of what constitutes a spacecraft and its component parts. The cost burden of a rescue mission is also not addressed in the agreement. The Rescue Agreement does provide that the launching state must bear the costs for the recovery of a craft that crashes into another state's territory.

However, the agreement makes no mention of the cost of the rescue of astronauts.

ADDITIONAL READING

Böckstiegel. (2002). *'Project 2001'-Legal Framework for the Commercial Uses of Outer Space*. Carl Heymanns Verlag. [*Institute of Air and Space Law*. McGill University.]. *Annals of Air and Space Law*, 2015–2020.

KEY TERMS AND DEFINITIONS

Mir: *Mir* (Russian: Мир, IPA: [ˈmʲir]; *lit.* peace or world) was a space station that operated in low Earth orbit from 1986 to 2001, operated by the Soviet Union and later by Russia. *Mir* was the first modular space station and was assembled in orbit from 1986 to 1996. It had a greater mass than any previous spacecraft. At the time it was the largest artificial satellite in orbit, succeeded by the International Space Station (ISS) after *Mir's* orbit decayed. The station served as a microgravity research laboratory in which crews conducted experiments in biology, human biology, physics, astronomy, meteorology, and spacecraft systems with a goal of developing technologies required for permanent occupation of space.

NASA: The National Aeronautics and Space Administration (NASA) is an independent agency of the U.S. federal government responsible for the civilian space program, as well as aeronautics and space research. NASA was established in 1958, succeeding the National Advisory Committee for Aeronautics (NACA). The new agency was to have a distinctly civilian orientation, encouraging peaceful applications in space science. Since its establishment, most US space exploration efforts have been led by NASA, including the Apollo Moon landing missions, the Skylab space station, and later the Space Shuttle. NASA is supporting the International Space Station and is overseeing the development of the Orion spacecraft, the Space Launch System, and Commercial Crew vehicles.

Space Shuttle Columbia Disaster: The Space Shuttle Columbia disaster was a fatal incident in the United States space program that occurred on February 1, 2003, when the Space Shuttle Columbia (OV-102) disintegrated as it reentered the atmosphere, killing all seven crew members. The disaster was the second fatal accident in the Space Shuttle program, after the 1986 breakup of Challenger soon after liftoff. During the launch of STS-107, *Columbia*'s 28th mission, a piece of foam insulation broke off from the Space Shuttle external tank and struck the left wing of the orbiter.

ENDNOTES

1 Piet-Hein Houben (1968), A New Chapter of Space Law: The Agreement on the Rescue and Return of Astronauts and Space Objects, Netherlands International Law Review, Vol.15, Issue2, June 1968, pp. 121-132. Retrieved from https://doi.org/10.1017/S0165070X00024049

2 https://en.wikipedia.org/wiki/Rescue_Agreement

Chapter 4
Main Contents and Comment on the 1972 Liability Convention

ABSTRACT

This chapter explains the historical background, key provisions, and main contents of the 1968 Space Liability Convention (Title: Convention on International Liability for Damage Caused by Space Objects). In order to protect the victims caused by space accidents, the author proposes a revision of the 1968 Liability Convention for the purpose of inserting and regulating the provision on the mental and indirect damage. The Convention on International Liability for Damage Caused by Space Objects, also known as the Space Liability Convention, is a treaty from 1972 that expands on the liability rules created in the Outer Space Treaty of 1967. In 1978, the crash of the nuclear-powered Soviet satellite Kosmos 954 in Canadian territory led to the only claim filed under the convention. The Liability Convention was concluded and opened for signature on 29 March 1972. It entered into force on 1 September 1972. As of 1 January 2019, 96 states have ratified the Liability Convention, 19 have signed but not ratified. States (countries) bear international responsibility for all space objects that are launched within their territory.

4.1. INTRODUCTION

The 1972 Liability Convention *(Title: The Convention on International Liability for Damage Caused by Space Objects)* is a treaty from 1972 that expands on the liability rules created in the Outer Space Treaty of 1967.

However, in 1978, the crash of the nuclear-powered Soviet satellite Cosmos 954 in Canadian territory nearly led to a claim under the Convention.

Skylab crashed in western Australia the following year, and while there were no deaths, injuries, or significant damage, the shire of Esperance did fine the Government of the United States $400 for littering as a result of Skylab's crashing to earth over Australia. NASA never paid the debt.

DOI: 10.4018/978-1-7998-7407-2.ch004

4.2. CONCLUSION OF THE LIABILITY CONVENTION

The Liability Convention was concluded and opened for signature in 1972. It entered into force in September 1972. As of 1 January 2019, 96 States including the USA, the United Kingdom, Canada, France, Germany, Russia, China, Japan, Italy, Spain, Brazil, Mexico, Australia, India, Indonesia, the Republic of Korea etc. have ratified the Liability Convention, 19 have signed but not ratified and four international (the European Space Agency, the European Organization for the Exploitation of Meteor Satellites, the Intersputnik International Organization of Space Communications, and the European Telecommunications Satellite Organizations) have declared their acceptance of the rights and obligations provided for in the Agreement.[1]

Figure 1.
Source: https://en.wikipedia.org/wiki/Space_Liability_Convention

4.3. KEY PROVISIONS

The basis of liability under the 1972 Space Liability Convention, and in particular examines fault-based liability. Both absolute liability and fault-based liability are found in Articles 2, 3, and 4 of the Space Liability Convention.:

Article II

A launching State shall be absolutely liable to pay compensation for damage caused by its space object on the surface of the earth or to aircraft flight. The Space Liability Convention forms the basis of liability for damage caused by a space object. There are three important aspects of this sentence that must be mentioned: “damage,” “space object,” and “caused by.”

Damage is defined as: "loss of life, personal injury or other impairment of health; or loss of or damage to property of States or of persons, natural or juridical, or property of international intergovernmental organizations."

This is a purposefully broad definition, as the type of damage flowing from space activities can lead to unpredictable effects, including environmental damage. Space object is defined as: "includes component parts of a space object as well as its launch vehicle and parts thereof."

This definition has spurred a great deal of academic debate due to the confusion that inevitably follows when a term is defined by itself: the definition in essence states that a space object is defined as a space object, its launch vehicle, and their parts.

However, the modifier "its" in "its launch vehicle" clearly references the term "space object," and as such we may presume that a space object must be an object that utilizes a launch vehicle. Further, it is not clear whether "space debris" is captured by the definition of space object, as either an independent space object or as a component part of a space object. There is no legal definition of space debris.

The Science and Technology Subcommittee of the United Nations Committee on the Peaceful Uses of Outer Space (UNCOPOUS) adopted a draft set of "high-level qualitative guidelines" in 2007,[2] which was subsequently endorsed by the United Nations General Assembly,[3] that defines space debris as: "All man-made objects including fragments and elements thereof, in Earth orbit or re-entering the atmosphere, that are non-functional."[4]

This definition is not legally binding, but "reflect[s] the existing practices as developed by a number of national and international organizations." The question of whether the pieces of a space object, when fragmented due to a collision, are still part of the original "space object" has yet to be addressed.

The phrase "caused by" is undefined within the Space Liability Convention. Clearly, the phrase references causation, though the applicable test for causation is not clear.

States (countries) bear international responsibility for all space objects that are launched within their territory. This means that regardless of who launches the space object, if it was launched from State A's territory, or from State A's facility, or if State A caused the launch to happen, then State A is fully liable for damages that result from that space object.

4.4. JOINT LAUNCHES

If two states work together to launch a space object, then both of those states are jointly and severally liable for the damage that object causes. This means that the injured party can sue either of the two states for the full amount of damage.

4.5. CLAIMS BETWEEN STATES ONLY

Claims under the Liability Convention must be brought by the state against a state.

The Convention was created to supplement existing and future national laws providing compensation to parties injured by space activities. Whereas under most national legal systems an individual or a corporation may bring a lawsuit against another individual or another corporation, under the Liability Convention claims must be brought on the state level only. This means that if an individual is injured by a space object and wishes to seek compensation under the Liability Convention, the individual must arrange for his or her country to make a claim against the country that launched the space object that caused the damage.

4.6. COMMENT

4.6.1. Liability Convention and Space Debris

Both space objects and space debris are carefully monitored by government agencies, and space objects are placed in orbits that are intended to avoid potential collisions with other space objects. But avoidance is not always possible due to the sheer amount of space debris. More than 21,000 orbital debris larger than 10 cm are known to exist. Who is liable, and how, if debris damages a satellite?

The estimated population of particles between 1 and 10 cm in diameter is approximately 500,000. The number of particles smaller than 1 cm exceeds 100 million. Even small pieces of debris can cause significant damage, as the average impact speed of space debris with a space object is 10 kilometers per second.

As a result, space objects must constantly analyze potential collisions and, if necessary, conduct avoidance procedures. The International Space Station must conduct such avoidance procedures to avoid collision approximately once per year. Despite these precautions, collisions do occur.

Figure 2. Sources of space debris include explosions of rocket bodies. Credit: ESA
Source: https://www.universetoday.com/134560/eye-opening-numbers-space-debris

On February 10, 2009, a Russian military satellite and private communications satellite owned by a US-based company collided in orbit. Both were destroyed.

The Russian satellite had been defunct since 1995, while the US satellite was still operational. The collision created at least 2,000 fragments of large space debris.[5]

No claim was made under the Space Liability Convention. The Liability Convention sets out a legal regime to provide compensation for damage caused in outer space by space objects. In outer space, liability is based on fault. It is significant to note that negotiations for the Liability Convention did not consider the question of the risks posed by space debris.

As a result, the negotiate did not address several liability issues of extreme importance related to damage caused by space debris. These issues include the meaning of *"damage"* and the reasonableness of a fault-based liability regime for damage caused in outer space by space debris. It would be possible to amend the Liability Convention so as to include damage to the outer space environment per se, based on the fact that outer space is a global commons.

The principle of fault-based liability is a further impediment to compensation for damage caused in outer space by orbital debris. Even if damage caused by space debris were within the scope of this regime, several other important legal issues, such as proof of negligence, gross negligence and contributory negligence, wilful-misconduct, among others, would remain unresolved.[6]

4.6.2. Jurisdiction and Control for Space Debris

Who has jurisdiction and control over space debris? It is significant to note that negotiations for the Liability Convention did not consider the question of the risks posed by space debris. It is a desirable thing for us to amend some articles of the 1972 Liability Convention after studying more deeply the said five items of it.

4.6.3. Indirect Damage

Though the Liability Convention regulated the damage caused by the space objects, but it does not regulated the extent of a causal relationship between space activity and effect occurrence as well as the nature and scope of damage. Generally, it is adequate to compensate for the *"direct damage"* without cause of parameters as a cause of the space activity.

However, it was a hot controversy for *"indirect damage"* relating to the protection of victims in those day of drafting for this Convention. Reading the Article 1 of the Convention it is a clear that only direct *'damage',* and not indirect damage, is contemplated by the Liability Convention. During the preliminary discussion several delegates had pointed out that the question of indirect damage could cause great difficulties in practice.

As a hypothesis example for the theory of damages in these neighborhoods as a result of the satellite debris fallen caused the heart attack, or a businessman's object to the space physical results from damage to the businessman and the financial institutions to borrow money in damages.

It was discussed also the issues for the consolation money such as "mental damages" and "*profit loss*" in concluding with the Liability Convention. Especially some countries has not regulated the compensation for damages on the mental loss and has enacted the punitive damage for wrongdoer in their domestic law. It was pointed out that the Liability Convention could not connected perfectly the protection of victims in the law of application.

ADDITIONAL READING

Kerr, S. (2017). *Liability for space debris collisions and the Kessler Syndrome*. Part 1. The Space Review.

Kim, D. H. (1994). Issues in International Air and Space Law, and in Commercial Law. Bobmun Publishing Company.

Kim, D. H. (2012). EU Draft Code of Conduct for Outer Space Activities: Space Debris and Liability Convention. In Decoding the International Code of Conduct for Outer Space Activities. Institute for Defense Studies & Analyses.

KEY TERMS AND DEFINITIONS

Contributory Negligence: Is the plaintiff's failure to exercise reasonable care for their safety. A plaintiff is the party who brings a case against another party (the defendant).

Indirect Damages: Means all indirect, consequential, special, incidental, punitive and aggravated damages and losses, loss of profits and diminution.

International Space Station (ISS): Is a modular space station (habitable artificial satellite) in low Earth orbit. It is a multinational collaborative project between five participating space agencies: NASA (United States), Roscosmos (Russia), JAXA (Japan), ESA (Europe), and CSA (Canada). The ownership and use of the space station is established by intergovernmental treaties and agreements. The station serves as a microgravity and space environment research laboratory in which scientific research is conducted in astrobiology, astronomy, meteorology, physics, and other fields. The ISS is suited for testing the spacecraft systems and equipment required for possible future long-duration missions to the Moon and Mars. Development and assembly of the station continues, with several major new Russian elements scheduled for launch starting in 2020. As of December 2018, the international space station is expected to operate until 2030.

Kosmos 954: (Russian: Космос 954) was a reconnaissance satellite launched by the Soviet Union in 1977. A malfunction prevented safe separation of its onboard nuclear reactor; when the satellite reentered the Earth's atmosphere the following year, it scattered radioactive debris over northern Canada, some of the debris landing in the Great Slave Lake next to Fort Resolution, NWT. This prompted an extensive cleanup operation known as Operation Morning Light. At 11:53 GMT on 24 January 1978, Kosmos 954 reentered the Earth's atmosphere while travelling on a northeastward track over western Canada. At first the USSR claimed that the satellite had been completely destroyed during re-entry, but later searches showed debris from the satellite had been deposited on Canadian territory along a 600-kilometre (370 mi) path from Great Slave Lake to Baker Lake.

Space Debris: Space debris, also called space junk, artificial material that is orbiting Earth but is no longer functional. This material can be as large as a discarded rocket stage or as small as a microscopic chip of paint. Much of the debris is in low Earth orbit, within 2,000 km (1,200 miles) of Earth's surface; however, some debris can be found in geostationary orbit 35,786 km (22,236 miles) above the Equator. As of 2020, the United States Space Surveillance Network was tracking more than 14,000 pieces of space debris larger than 10 cm (4 inches) across. It is estimated that there are about 200,000 pieces between 1 and 10 cm (0.4 and 4 inches) across and that there could be millions of pieces smaller than 1 cm. How long a piece of space debris takes to fall back to Earth depends on its altitude. Objects below 600 km (375 miles) orbit several years before reentering Earth's atmosphere. Objects above 1,000 km (600 miles) orbit for centuries.

Willful Misconduct: Generally means a knowing violation of a reasonable and uniformly enforced rule or policy. It means intentionally doing that which should not be done or intentionally failing to do that which should be done, knowing that injury to a person will probably result or recklessly disregarding the possibility that injury to a person may result. The term is applied in various legal contexts, such as employment and, torts, and public offices.

ENDNOTES

1 https://en.wikipedia.org/wiki/Liability_Convention

2 Space Debris Mitigation Guidelines of the Committee on the Peaceful Uses of Outer Space, as annexed to UN doc. A/62/20 (2007), Report of the COPUOS.

3 United Nations General Assembly, Resolution 62/217 on the international cooperation in the peaceful uses of outer space, (2008 62nd session.

4 Space Debris Mitigation Guidelines of the Committee on the Peaceful Uses of Outer Space, as annexed to UN doc. A/62/20, Report of the COPUOS (2007) at 1.

5 National Aeronautics and Space Administration (NASA), "Orbital Debris Quarterly News" (2011), Orbital Debris Program Office.

6 https://www.boulder.swri.edu/NSRC2020/Site5/Home.html

Chapter 5
Main Contents and Comment on the 1975 Registration Convention

ABSTRACT

This chapter describes the 1975 Registration Convention including the key provisions and main contents of the Registration Convention (Title: The Convention on Registration of Launched Objects Into Outer Space). Although currently there are 69 parties to the REGISTRATION Convention, many lawyers, professors, scientists, etc. urge states that have not yet become parties to the international treaties governing the uses of outer space to give consideration to ratify or accede to those treaties in accordance with their national law, as well as incorporating them into their national legislation. The 1976 Registration Convention, or Convention on Registration of Objects Launched Into Outer Space, which obligates parties to register launches of all objects launched into Earth orbit or into outer space with an appropriate national space agency, was considered and negotiated by the COPUOS Legal Subcommittee from 1962. It was adopted by the General Assembly in 1974, General Assembly Resolution 3235 (XXIX); opened for signature on January 14, 1975; and entered into force on September 15, 1976.

5.1. INTRODUCTION

The 1976 Registration Convention (Title: *Convention on Registration of Objects Launched into Outer Space*), which obligates Parties to register launches of all objects launched into Earth orbit or into outer space with an appropriate national space agency, was considered and negotiated by the COPUOS *Legal Subcommittee* from 1962.

It was adopted by the General Assembly in 1974, General Assembly Resolution 3235 (XXIX), opened for signature on January 14, 1975 and entered into force on September 15, 1976.

DOI: 10.4018/978-1-7998-7407-2.ch005

As of December 2018, it has been ratified by 70 states including the USA, the United Kingdom, Canada, France, Germany, Russia, China, Japan, Italy Mexico, Brazil, Argentina, Australia, Spain, Mexico, India, Indonesia, the Republic of Korea and North Korea etc.[1] The convention requires states to furnish to the United Nations with details about the orbit of each space object.

This 1976 Registration Convention, drawn up by the United Nations Committee on the Peaceful Uses of Outer Space (COPUOS), was the fourth treaty exclusively dedicated to outer space.

The text of the 1976 Registration Convention was, after some years of discussion and drafting, adopted in New York on November 12, 1974, and made public through a Resolution of the General Assembly, Resolution 3235(XXIX), which contained the text in the Annex and opened the Registration Convention for signature as per January 14, 1975.

It entered into force quite rapidly on September 15, 1976, after the fifth instrument of deposit had been received with the Secretary-General of the United Nations.

Building upon the desire expressed by States in the 1967 Outer Space Treaty, the 1968 Rescue Agreement and the 1972 Liability Convention to make provision for a mechanism that provided States with a means to assist in the identification of space objects, the 1976 Registration Convention expanded the scope of the United Nations Register of Objects Launched into Outer Space that had been established by *the UN Resolution 1721B (XVI) of December 1961* and addressed issues relating to States Parties responsibilities concerning their space objects.

The Secretary-General was, once again, requested to maintain the Register and ensure full and open access to the information provided by States and international intergovernmental organizations.

Since 1962, the United Nations has maintained a *Register of Objects Launched into Outer Space.* Originally established as a mechanism to aid the United Nations Committee on the Peaceful Uses of Outer Space (COPUOS) in its discussions on the political, legal and technical issues concerning outer space, the evolution of international space law resulted in space object registration becoming a means of identifying which States' bear international responsibility and liability for space objects.

Following multi-year discussion among States, the 1976 Registration Convention or *Convention on Registration of Objects Launched into Outer Space* entered into force in 1976. States and international intergovernmental organizations that agree to abide by the Convention are required to establish their own national registries and provide information on their space objects to the Secretary-General for inclusion in the United Nations Register.

Responsibility for maintenance of the Register was delegated by the Secretary-General to the *United Nations Office for Outer Space Affairs* (UNOOSA). As required under the treaty, UNOOSA publicly disseminates the information provided as United Nations documents, which are available through its website and through the United Nations Official Document System (ODS). To date over 88% of all satellites, probes, landers, crewed spacecraft and space station flight elements launched into Earth orbit or beyond have been registered with the Secretary-General.[2]

The Registration Convention and four other space law treaties are administered by the United Nations Committee on the Peaceful Uses of Outer Space. The European Space Agency and European Organization for the Exploitation of Meteorological Satellites have submitted declarations of acceptance of rights and obligations according to the convention.

5.2. CURRENT STATUS

The register is kept by the United Nations Office for Outer Space Affairs (UNOOSA) and includes.

- Name of launching State
- An appropriate designator of the space object or its registration number
- Date and territory or location of launch
- Basic orbital parameters (Nodal period, Inclination, Apogee and Perigee)
- General function of the space object

Information on registered objects is available at the UNOOSA website. It is then the responsibility of the UN Secretary-General to maintain a public UN Register of Objects Launched into Outer Space (Article III), based on the information provided by States. Under Article IV (1) of the Registration Convention, each State of registry is required to provide to the UN Secretary-General, as soon as practicable, specified information concerning each space object carried on its registry (the name of launching State or States, an appropriate designator of the space object or its registration number, the date and territory or location of launch; basic orbital parameters).

5.3. COMMENT

Currently there are 69 Parties to the Registration Convention (including for example, Canada, Australia, the US and the United Kingdom); and, the annual UNGA resolution on International Cooperation in the Peaceful Use of Outer Space (most recently resolution 69/85 of 5 December 2014) "urges States that have not yet become parties to the international treaties governing the uses of outer space to give consideration to ratifying or acceding to those treaties in accordance with their national law, as well as incorporating them into their national legislation".

ADDITIONAL READING

Jakhu, R., & Pelton, J. (Eds.). (2017). Global Space Governance: An International Study. Springer.

McCarthy, K. (2015). *U.S. Commercial Space Launch Competitiveness Act.* Act No. H.R.2262 of 5 December 2015. 114th Congress (2015-2016).

KEY TERMS AND DEFINITIONS

European Space Agency: The European Space Agency (ESA) is an inter-governmental organization of 22 member states dedicated to the exploration of space. Established in 1975 and headquartered in Paris, ESA has a worldwide staff of about 2,200 in 2018 and an annual budget of about €6.68 billion (US$7.43 billion) in 2020. ESA's space flight programme includes human spaceflight (mainly through participation in the International Space Station program); the launch and operation of un-crewed exploration missions to other planets and the Moon; Earth observation, science and telecommunication; designing launch vehicles; and maintaining a major spaceport, the Guiana Space Centre at Kourou, French Guiana.

Launching State: "Launching State" literally means the state that, under its command, a rocket carrying any kind of space object on board has been launched into the Earth's orbit or beyond. Although the space law has added other supplementary explanations into this initial definition that will be scrutinized in further lines. The main objective of addressing the term of "launching state" in the body of space law documents has been to present straightforward as well as practical rules for easily identifying the state that in the event of any accident should be taken liable to remedy the damage and loses caused by the space object which it had previously launched into space.

Space Object: The term *Space Object* in reference to outer space was first used in 1961 in General Assembly Resolution 1721 (XVI) titled *International cooperation in the peaceful uses of outer space* to describe any object launched by States into outer space. Professor Bin Cheng, a world authority on International Air and Space Law, has noted that members of the COPUOS during negotiations over the space treaties treated spacecraft and space vehicles as synonymous terms. The Space Object can be considered as the conventional launcher, the reusable launcher, the satellite, the orbital station, the probe, the impactor, the space telescope.

ENDNOTES

1 https://en.wikipedia.org/wiki/Registration_Convention
2 https://www.spacelegalissues.com/the-1976-registration-convention/

Chapter 6
Main Contents and Comment on the 1979 Moon Agreement of 1979

ABSTRACT

This chapter explains the ratification, main contents, and prospect of the 1979 Moon Agreements. The Agreement Governing the Activities of States on the Moon and Other Celestial Bodies, better known as the Moon Treaty or Moon Agreement, is a multilateral treaty that turns jurisdiction of all celestial bodies (including the orbits around such bodies) over to the participant countries. Thus, all activities would conform to international law, including the United Nations Charter. It has not been ratified by any state that engages in self-launched human spaceflight or has plans to do so (e.g., the United States, the larger part of the member states of the European Space Agency, Russia [former Soviet Union], People's Republic of China, and Japan). As of January 2019, 18 states are parties to the treaty. As the current Moon Agreement has emerged as a problem as the United States and other major powers are not joining it, many lawyers, professors, and scientists urged that the powers ratify it quickly.

6.1. INTRODUCTION

The *Agreement Governing the Activities of States on the Moon and Other Celestial Bodies,* better known as the Moon Agreement, is an international treaty that turns jurisdiction of all celestial bodies (including the orbits around such bodies) over to the international community. Thus, all activities must conform to international law (notably this includes the UN Charter).

In practice it is a failed treaty since it has not been ratified by any nation which engages in self-launched manned space exploration or has plans to do so (e.g. the United States, some member states of the European Space Agency, Russian Federation, People's Republic of China, Japan, India and the Republic of Korea etc.) since its creation in 1979, and thus has a negligible effect on actual space-flight.[1] After ten more years of negotiations, the Moon Agreement was created in 1979 as a framework of laws to develop a regime of detailed procedures.

DOI: 10.4018/978-1-7998-7407-2.ch006

6.2. RATIFICATION

The Moon Agreement of 1979 and, after satisfying the condition requiring five ratifying states, it entered into force for the ratifying parties in 1984. As of January 2019, 18 states (Armenia, Australia, Austria, Belgium, Chile, Kazakhstan, Kuwait, Lebanon, Mexico, Morocco, Netherlands, Pakistan, Peru, Philippines, Saudi Arabia, Turkey, Uruguay and Venezuela) are ratified and acceded.

Four additional states, France, Guatemala, India, Romanian have signed but not ratified the treaty. The Moon Agreement opposed ratification of the treaty by the United States Senate.

The last effort culminated in June 2018 after eight years of negotiations, when the United Nations Committee on the Peaceful Uses of Outer Space (COPUOS) held a high-level meeting in Vienna, Austria that tried to produce a consensus on a framework of laws for the sustainable development of outer space, but it also failed to do so[2]when S. Neil Hosenball, who is the NASA General Counsel and chief US negotiator for the Moon Agreement, decided that negotiation of the rules of the international regime should be delayed until the feasibility of exploitation of lunar resources has been established.[3]

If rights to economic benefits cannot be guaranteed, there will be little if any private investment.[4] So seeking clearer regulatory conditions and guidelines,[5v] private companies in the US prompted the US government and legalized space mining in 2015 by introducing the US Commercial Space Launch Competitiveness Act of 2015.[6] Similar national legislations legalizing extraterrestrial appropriation of resources are now being replicated by other nations, including Luxembourg etc. [7]

Figure 1.

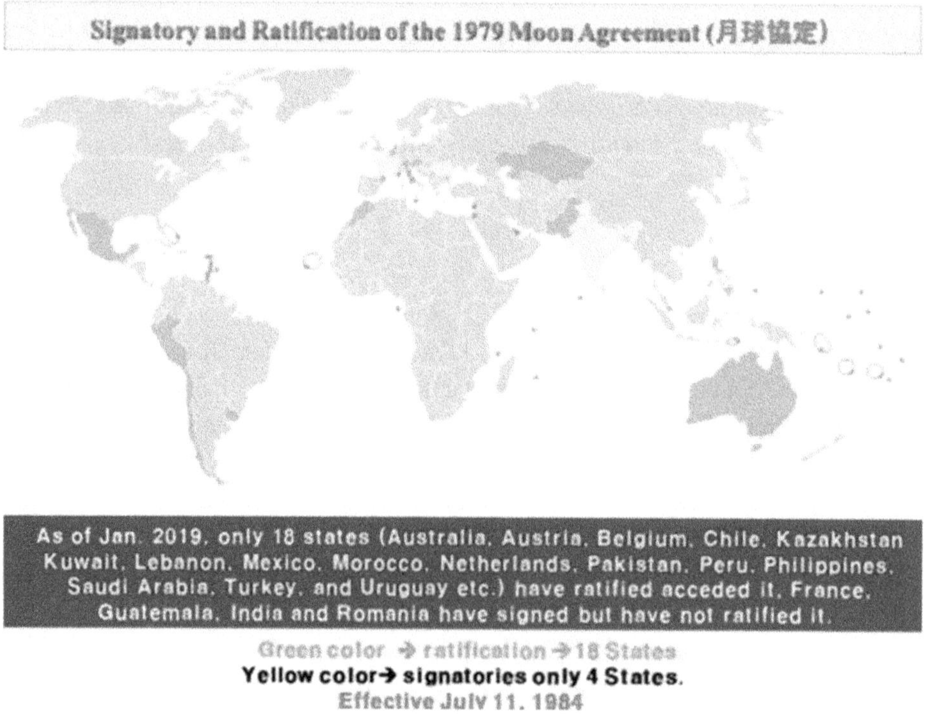

6.3. NEGOTIATIONS BY UNCOPUOS

While the "national" treaty explicitly allows commercial mining, other experts argue that these new national laws are inconsistent with the Moon Agreement and customary international law.[8] Other experts affirm that the Moon Agreement does allow commercial mining, after creating the needed international regulations. This has created a controversy on claims and on mining rights.

6.4. MAIN CONTENTS

The Moon Agreement would apply to the moon and to other celestial bodies within the Solar System, other than the Earth, including orbits around or other trajectories to or around them. It also expresses a desire to prevent the Moon from becoming a source of international conflict. The Moon Agreement proposes to establish an "international regime" or "framework of laws" that apply to the Moon and to other celestial bodies within the Solar System, including orbits around or other trajectories to or around them.[9]

The Moon Agreement lays several provisions outlined in 21 articles. In Article 1, the treaty makes a declaration that the Moon should be used for the benefit of all states and all peoples of the international community. It reiterates that lunar resources are "not subject to national appropriation by claim of sovereignty, by means of use or occupation, or by any other means."[10]

It also expresses a desire to prevent the Moon from becoming a source of international conflict, so that the resources should be used exclusively for peaceful purposes. To those ends, the treaty lays several provisions, and some of these are paraphrased below:

- Bans any military use of celestial bodies, including weapon testing, nuclear weapons in orbit, or military bases. The use of military personnel for scientific research or for any other peaceful purposes shall not be prohibited (Article 3.4).
- Provides a framework of laws to establish an international cooperation regime, including appropriate procedures, to govern the responsible exploitation of natural resources of the Moon (Article 11.5).
- Bans altering the environmental balance of celestial bodies and requires that states take measures to prevent accidental contamination of the environments of celestial bodies, including Earth (Article 7.1).
- The orderly and safe use of the natural lunar resources with an equitable sharing by all state parties in the benefits derived from those resources (Article 11.7).
- The moon and its natural resources are the common heritage of mankind (Article 11.1).
- There shall be freedom of scientific research and exploration and use on the Moon by any party without discrimination of any kind (Article 6). Samples obtained during research activities, are hoped to be made available to all countries and scientific communities for research (Article 6.2).
- Shall promptly inform the United Nations and the public of any phenomena which could endanger human life or health, as well as of any indication of extraterrestrial life (Article 5.3).
- State parties shall ensure that non-governmental entities under their jurisdiction shall engage in activities on the Moon only under the authority and continuing supervision of the appropriate state party (Article 14).

- All parties shall inform the United Nations as well as the public, of their activities concerned with the exploration and use of the Moon (Article 5).
- Any state party to this agreement may propose amendments to the agreement. (Article 17).

When compared with the Outer Space Treaty, it reiterates most provisions, and adds two new concepts in order to address the exploitation of natural resources in outer space: to apply the concept of 'common heritage of mankind' to outer space activities, and to have the participating countries produce a regime that lays the appropriate procedures for orderly mining. Multiple conferences produced no consensus on these two items.

6.5. LEGAL PROBLEMS

The spacecraft flew over the northern regions of the moon on 7 December 1992.

The colors indicate different materials. The current imprecision of the Moon Agreement generated various interpretations,[11] and it is cited as the main reason it was not ratified by most parties.[12] Only one country (India) with independent spaceflight capabilities has signed (but not ratified) the treaty.

An expert in space law and economics thinks that the treaty would need to offer adequate provisions against any one company acquiring a monopoly position in the world minerals market, while avoiding "the socialization of the Moon." Another expert commends the treaty as a germinal legal framework for developing the needed laws, rather than a finished set of detailed laws.[13]While the treaty reiterates the prohibition of sovereignty of "any part" of space, it proposes that the exploitation of resources shall be governed by an international regime (Article 11.5), but there has been no consensus establishing these laws S. Neil Hosenball, who is the NASA General Counsel and chief US negotiator for the Moon Treaty, decided in 2018 that negotiation of the rules of this international regime should be delayed until the feasibility of exploitation of lunar resources has been established.

A legal expert stated in 2020 that the international issues "would probably be settled during the normal course of space exploration."

6.6. COMMENT

Recently it is most severe competition among the space superpowers in order to mine and exploit the natural resources including Helium-3 from the moon so as to solve the serious problems of the earth's energy. As it is un-ratified by any major space-faring powers and unsigned by most of them, it is of no direct relevance to current space activities. The space superpowers and private operators have not started to exploit the natural resources of the Moon and other celestial bodies yet is the absence of rules setting out how this exploitation shall be carried out.

The space law system, indeed, does not provide any specific rule, relating to the exploitation and development of extraterrestrial natural resources, which have been generally accepted by States. According to the 1967 Outer Space Treaty (OST) and the 1979 Moon Agreement, these two instruments does not offer an adequate legal framework which is able to ensure the safe, orderly and peaceful development and extraction and development from the natural resources of the Moon and other celestial bodies.

On one side, the Outer Space Treaty does not contain any mention of space resources or to their possible exploitation and development of minerals in the moon such as mining right etc. On the other side, the Moon Agreement, whose main purpose is to set forth rules aimed at regulating the use for scientific and commercial reasons of lunar and other celestial bodies' materials, has been rejected by the majority of States, comprising the space-faring States.[14]As a consequence, its principles lose relevance when applied to the exploitation of extraterrestrial resources. The major problem of the Treaty is that it does not contain any specific reference to space resources and to their exploitation. The exploitation of moon materials raises several specific legal issues, such as those related to the right of mining extraterrestrial sites or to property rights over the extracted materials, which may not be properly dealt with and solved by simply relying on the existing space law principles.

It is realistic to anticipate that the exploitation of these mineral resources in moon will take place in a three phase process;

1. pre mining phase;
2. mining phase;
3. post mining phase.[15]

The rules regulating the exploitation of the natural resources of the moon and other celestial bodies should be inserted in a legal instrument which will be opened for acceptance by State and International Organizations. As the 1967 Outer Space Treaty and 1979 Moon Agreement has not legal provisions for the development of many natural resources such as Helium 3, iron, manganese, silicon, titanium, etc., so it is necessary for us to revise the aforementioned two treaties in order to legislate the new legal provisions including the license, essence and duration of the mining right in the near future.

The International Space Agency (ISA: tentative name) based on Article 11, 5~8 and Article 18 of Moon Agreement has to establish necessary as the international organization that States manage and control the exploitation and development of the natural resources in the moon and other celestial bodies. The ISA has the power to authorize persons to exploit for commercial purposes a certain natural resources in the moon or Other celestial bodies. At the same time, however, the ISA has the duty to control that the mining activities for minerals are carried out in accordance with the space law principles and in a not detrimental manner for the space environment.

ADDITIONAL READING

O'Brien, D. C. (2019). *Beyond UNISPACE: It's time for the Moon Treaty*. The Space Review.

Van Bogaert, E. R. C. (1986). *Aspects of Space Law*. Kluwer.

KEY TERMS AND DEFINITIONS

Helium-3: Existence of lunar helium-3, to be used as fuel for fusion reactors, is well documented; verified from numerous Apollo and Luna mission samples, current analyses indicate that there are at least 1 million tones embedded in the lunar surface. The Helium-3 would be used as fuel for fusion reactors.

Moon gas may solve earth's energy crisis. The value of Helium-3 is that it can generate nuclear power and, as a consequence, energy in a clean way, namely through a process of nuclear fusion which does not produce toxic waste. Thanks to these special characteristics, the extraction of Helium-3 is likely to have a huge impact on the way energy is produced and distributed on earth. Helium-3 is deposited on the lunar surface by solar winds and would have to be extracted from moon soil and rocks. As space superpowers such as the United States, Russia, European Union, China, Japan and India has interested in Helium-3 that it is more expensive 300 times than gold, so it may be become a supremacy country in the future resources war from mining Helium-3 in the moon to bring it to earth for the purpose of getting it in advance.

Mining Right: Mining Right means an exclusive license or licenses to commercially mine the gold and other metals or minerals in the Project Area. Mining rights are property rights to exploit and develop an area for the minerals it harbors. Mining rights are the right to dig and acquire registered minerals and other minerals existing in similar deposits in certain land areas (mining areas) registered under the Mining Act. The detailed provisions of mining rights are stipulated in the Mining Law. Mineral rights can refer to sedentary minerals that do not move below the Earth's surface or fluid minerals such as oil or natural gas. There are two types of mining rights such as exploration right and mineral rights.

Natural Lunar Resources: The Moon bears substantial natural resources which could be exploited in the future. Potential lunar resources may encompass process able materials such as volatiles and minerals, along with geologic structures such as lava tubes that together, might enable lunar habitation. The use of resources on the Moon may provide a means of reducing the cost and risk of lunar exploration and beyond. Solar power, oxygen, and metals are abundant resources on the Moon. Elements known to be present on the lunar surface include, among others, hydrogen (H), oxygen (O), Silicon (Si), iron (Fe), magnesium (Mg), calcium (Ca), aluminum (Al), manganese (Mn) and titanium (Ti).

Other Celestial Bodies: Celestial bodies or heavenly bodies are objects in space such as the sun, moon, planets and stars. A natural celestial object, visible in the sky, such as a star, planet, natural satellite, asteroid, comet, the Moon or the Sun. The term celestial body is as expansive as the entire universe, both known and unknown. By definition a celestial body is any natural body outside of the Earth's atmosphere. Easy examples are the Moon, Sun, and the other planets of our solar system. But those are very limited examples. The Kuiper belt contains many celestial bodies. Any asteroid in space is a celestial body.

Solar System: The Solar System is the gravitationally bound system of the Sun and the objects that orbit it, either directly or indirectly. Of the objects that orbit the Sun directly, the largest are the eight planets, with the remainder being smaller objects, the dwarf planets, and small Solar System bodies. Of the objects that orbit the Sun indirectly—the moons—two are larger than the smallest planet, Mercury. The Solar System also contains smaller objects. The asteroid belt, which lies between the orbits of Mars and Jupiter, mostly contains objects composed, like the terrestrial planets, of rock and metal. The vast majority of the system's mass is in the Sun, with the majority of the remaining mass contained in Jupiter. The four smaller inner planets, Mercury, Venus, Earth and Mars, are terrestrial planets, being primarily composed of rock and metal.

ENDNOTES

1 https://en.wikipedia.org/wiki/Moon_Treaty

2 U.S. Congress. Senate, Subcommittee on Science, Technology, and Space of the Committee of Commerce, Science, and Transportation. *Hearings before Subcommittee on Science, Technology, and Space of the Committee of Commerce, Science, and Transportation*, 96th Congress, 2nd sess., 1980.

3 Beldavs, Vidvuds (15 January 2018). *"Simply fix the Moon Treaty"*. The Space Review.

4 O'Brien, Dennis (5 March 2018). *"Why it's a bad idea to weaken the Moon Treaty"*. Space Review. [...] the Moon Treaty, must be revised so that investors in a future space economy can achieve a sustainable return on their investments.

5 *Why it's a bad idea to weaken the Moon Treaty.* Dennis O'Brien, *The Space Review*. 5 March 2018. Quote: "Its not really so much what the regulation is but that there is regulatory clarity. Right now, because there are no clear regulations on these activities, companies are afraid, [...]"

6 H.R.2262 - U.S. Commercial Space Launch Competitiveness Act. 114th Congress (2015-2016). Sponsor: Rep. McCarthy, Kevin. 5 December 2015.

7 *If space is 'the province of mankind', who owns its resources?* Senjuti Mallick and Rajeswari Pillai Rajagopalan. *The Observer Research Foundation*. 24 January 2019.

8 Ridderhof, R. (18 December 2015). *"Space Mining and (U.S.) Space Law"*. Peace Palace Library. Retrieved 26 February 2019.

9 Agreement Governing the Activities of States on the Moon and Other Celestial Bodies. - Resolution 34/68 Adopted by the General Assembly. 89th plenary meeting; 5 December 1979.

10 Schingler, J. K.; Kapoglou, A. (July 15–17, 2019). "Common Pool Lunar Resources" (PDF). Lunar ISRU 2019: Developing a New Space Economy Through Lunar Resources and Their Utilization. Columbia, Maryland.

11 Schingler, J. K.; Kapoglou, A. (July 15–17, 2019). *"Common Pool Lunar Resources" (PDF). Lunar* ISRU 2019: Developing a New Space Economy Through Lunar Resources and Their Utilization. Columbia, Maryland.

12 Extraterrestrial Property and Space Law: Frequently Asked Questions. *Moon Estates*. Accessed on 6 November 2019.

13 The 1979 Moon Agreement. Louis de Gouyon Matignon, *Space Legal Issues*. 17 July 2019.

14 The refusal of the developed States to ratify the Moon Agreement was largely due to the insertion of the Common Heritage of Mankind idea in Article XI of the Agreement declaring the Moon and its resources to be "the Common Heritage of Mankind", see C.Q.Christol, "*Important concepts for international law of outer space*", in Proceedings of the Fortieth Colloquium on the Law of Outer Space, (1997), p. 73; F.G.von de Dunk, "*The dark side of the Moon: public concepts and private enterprises*", in Proceedings of the Fortieth Colloquium on the Law of Outer Space, (1997), p. 121.

15 Fabio Tronchetti, *op.cit.*, pp. 133~144.

Chapter 7
The Asian Countries' National Legislation on the Outer Space Law and Organizations

ABSTRACT

This chapter looks at Asian countries and their legislation on outer space laws.

7.1. CHINA

7.1.1. The 2001 Measures for the Administration of Registration of Objects Launched Into Outer Space

China has enacted two 'ministerial' regulations in order to implement their registration and licensing obligations under the Space-Specific treaties, these are:

- *Measures for the Administration of Registration of Objects Launched into Outer Space 2001* (CHN) (*'Registration Measures*'); and
- *Interim Measures on the Administration of Licensing the Project of Launching Civil Space 2002* (CHN)('Licensing Measures').

Ministerial regulations are considered some of the lowest priority rules in the Chinese legal system. As such, the two measures highlight China's lack of emphasis on the development of their space legal regime as opposed to their significant efforts toward their space program itself.[1]

The purpose of the 2001 Registration Measures is to ensure the implementation of the requirements set in the 1975 Registration Convention, which demands states parties to register their space objects in a national register as well as to transfer relevant information to the U.N. Secretary-General for inclusion in an international registry.

DOI: 10.4018/978-1-7998-7407-2.ch007

The Registration Measures contain 16 articles addressing several aspects related to the registration of space objects.

They begin by defining a space object as "an artificial satellite, crewed spacecraft, space probe, space station, launch vehicle and parts thereof, and other human-made objects launched into outer space."

The *Licensing Measures 2001* (CHN) were implemented in order to provide a regime for licensing civil space launches in China. China is internationally responsible for its national activities under Article IV of the *Outer Space Treaty*, meaning it is under an obligation to authorize and supervise the activities of its entities. Space license application form with attachments outlining the applicant's relevant qualifications and documentation or evidence that the project complies with national laws and regulations of China. A permit holder must apply to SASTIND at least 90 days prior to the planned launch date in order to cancel or modify a valid permit.

The *Registration Measures 2001* (CHN) were largely enacted in order to fulfil China's obligations under the 1975 *Registration Convention.*

Where China launches or procures the launch of a space object it must register the space object on its own national register, as well as the United Nations register maintained by the UN Secretary-General.

Whilst the *Registration Measures 2001* (CHN) are essentially a restatement of the principles set out under the *Registration Convention*, China employs a broader definition of the term 'space object' than that under the *Convention.*[2]

The national register is currently maintained by SASTIND through the Chinese National Space Administration (CNSA: 中国国家航天局). As such, all space objects launched within China, or launched jointly by China, are to be registered with the State Administration for Science, Technology and Industry for National Defense (SASTIND; Chinese: 国家国防科技工业局) within 60 days after their launch into orbit. Overall, the Registration Measures represent a valid instrument to implement the requirements set forth in the Registration Convention, at times even improving upon them.

Four big satellite and spacecraft launch center is located at Jichuan(酒泉), Taiyuan (太原), Xichang (西昌), Wenchang (文昌) in China. However, it is also true that the Measures' provisions present some shortcomings.

Figure 1.
Sources: https://www.sohu.com/a/155445477_249601

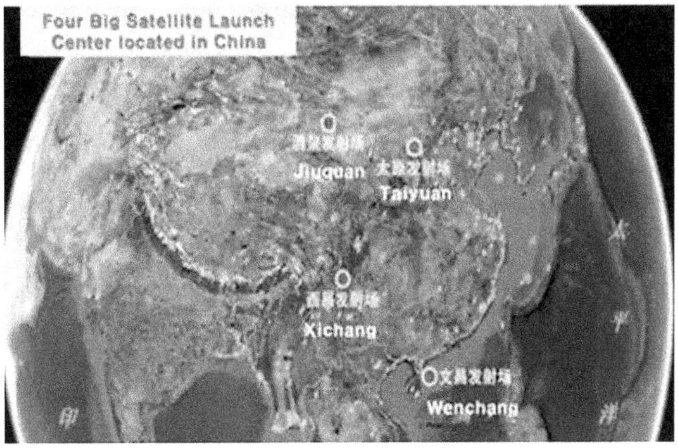

For example, while referring to objects launched into outer space, they do not clarify where outer space begins; this may become an issue in relation to the applicability of the Measures to objects that operate at the border between airspace and outer space, namely at altitudes between 80 and 100 km above sea level.

Another problem concerns the interpretation of the expression "main owner" of a space object, as neither the Registration Measures nor the Registration Convention use this expression.[3] The purpose of the 2002 Licensing Measures is to set forth the legal regime for the licensing of launches of civil, non-military, objects into outer space.

The Measures include five chapters and 26 articles that lay down detailed procedures on how to apply for a license, on one side, and describe the rights and duties of the licensee, the supervisory mechanisms and the penalties to be imposed in case of violation of the license's terms, on the other side.

7.1.2. The 2002 Interim Measures on the Administration of Permits for Civil Space Launch Projects

The Licensing Measures are applicable to the launching of space objects from Chinese territory as well as to the launching of space objects from foreign territory if the space object is owned by China or if its ownership is transferred to Chinese natural or juridical persons or organizations.

Thus, any person or organization willing to undertake a launch project shall apply for examination and approval from the State Administration for Science, Technology and Industry for National Defense (SASTIND; 国家国防科技工业局), and it is prohibited from carrying such project until an authorization is obtained.

The need for such an authorization is to be found in Article VI of the Outer Space Treaty, which establishes the international responsibility of states for national activities in outer space, including those of governmental and non-governmental entities, and requires states to authorize and continuously supervise the activities of non-governmental entities. The general project contractor or the final owner of the satellite or other spacecraft are under the obligation to apply for the license. Such a license can be obtained if the proposed activity:

1. does not endanger national security and public health;
2. is consistent with environmental protection laws;
3. and the applicant possess the required technical and financial capabilities.

The applicant is further requested to obtain an insurance covering liability for damage or other losses caused to third parties and other liability cases incurred by launching a space object. Applications shall be submitted to the SASTIND nine months prior to the launch.

The SASTIND shall review the application within 30 days from receipt and shall grant or reject the attribution of a license. Once issued, a license shall include the name and address of the licensee, the main elements of the project, the time frame of the launch and the duration of the project. A license cannot be altered or transferred and it shall be immediately terminated once the project is completed.

If the licensee wishes to modify information in the license or cancel it, the SASTIND should be notified 90 days before the expiration of the license for modification or cancellation.

The licensee shall comply with the terms of the license and with the laws and regulations of China. In case of violation, the SASTIND shall order the licensee to rectify the violation within a deadline or shall withdraw the license if the violation is serious. Overall, the Licensing Measures provide a good foundation to comply with the requirements of Articles VI of the Outer Space Treaty and to administer the launch of civil space objects.

However, there are two aspects of the Measures that remain controversial. First, it remains doubtful whether the scope of the license, apart from covering the launch itself, also extends to activities actually occurring in outer space once the launch is complete.

In other words, as the Measures only focus on the "launching" phase, it is not clear whether the behavior of the licensee in space is regulated by the Measures and if the government is provided with effective means to control it.[4]

7.1.3. Chinese Legislation on the Space Law in the Near Future

There is no space law in China meaning the law is disproportionally under-developed in comparison with China's space capabilities and legislation in other nations. A mixture of historical, political, legal, and bureaucratic reasons accounts for this absence. Space law was incorporated into the 5-year legislation plan of National People's Congress of China in 2013. China will always abide by international space law and is making great efforts to introduce its own laws in 2020, the China National Space Administration (CNSA) said on Monday. Xu Dazhe, director of the CNSA, said the country has already made policies and regulations in the administration of civil space launch, registration of space objects and reduction and prevention of space debris, at the 2014 Workshop on Space Law.

"National space law has been listed in the national legislation plan, and the CNSA is drafting the law," said Tian Yulong, secretary-general of the CNSA.[5]

Simonetta Di Pippo, director of the United Nations Office for Outer Space Affairs (UN-OOSA) said the workshop will help each country abide by the international space law, make better use of outer space and maintain space safety. She said that the workshop will cover the developments in space law and policy, space law and commercial space activities, promoting national space legislation and space capacity and information.[6]

Attended by experts from more than 30 countries and international organizations, the workshop is co-organized by the UN-OOSA, the CNSA and the Asia-Pacific Space Cooperation Organization. China has made rapid development in space exploration and now rushing to establish basic infrastructure in outer space, including remote sensing, communication and navigation.

In the last decade China has achieved a remarkable level of success in the space sector and has rapidly become one of the most dynamic and innovative space players. China does not have a comprehensive, fully fledged national space law. Instead, Chinese space activities are run through a number of internal management rules and departmental regulations issued by space authorities.

Mr. Xu Dazhe (许达哲), Director of the China National Space Administration (CNSA: 中国国家航天局)[7] said that China has already made policies and regulations in the administration of civil space launch, registration of space objects and reduction and prevention of space debris. "National space law has been listed in the national legislation plan, and the CNSA is drafting the law," said Mr. Tian Yulong, Secretary-General of the CNSA. China expects to introduce space law around 2020.

The first draft of China's "Space Law" preparation work was completed by the aroused of December, 2016.

Its purpose is to regulate the activities of internal space, external to promote the implementation of international duties and obligations, and through legal channels to regulate all aspects. It is understood that the law has been included in the legislative plan of the State Council, the preparation of the draft has been prepared for four or five years.

The document outlines rules for the research, development, testing of launch vehicles, safety, confidentiality and export control, interaction with launch sites, dealing with propellants, as well as listing supporting laws and regulations for China's space activities. The rules also clarify what qualifications are required by commercial aerospace enterprises, the scope of business, as well as what support may be obtained from the government, and underlines the role of the civil-military integration national strategy in fostering development in the sector.

The Chinese space regulations were jointly developed by the State Administration of Science, Technology and Industry for National Defense (SASTIND), which oversees the country's space activities, and the Equipment Development Department of the Central Military Commission and announced June 10, 2019.

The formulation of guidelines has largely been a top-down process, with no indication of involvement of commercial space actors. A wider, first national space has been included in the legislative plan of China's parliament, the National People's Congress, and could be introduced before 2023.[8]

7.1.4. China: Chang'e 4, Late 2019, 5 and 6, Late 2020 and 2023

Chang'e 4 (Chinese: 嫦娥四号) is a robotic spacecraft mission, part of the second phase of the Chinese Lunar Exploration Program. China achieved humanity's first soft landing on the far side of the Moon, on 3 January 2019. A communication relay satellite, Queqiao, was first launched to a halo orbit near the Earth–Moon L2 point in May 2018.

The robotic lander and Yutu-2 (Chinese: 玉兔二号) rover were launched on December 7, 2018 and entered lunar orbit on December 12, 2018, before landing on the Moon's far side. The mission is the follow-up to Chang'e 3, the first Chinese landing on the Moon.[9] In early 2019, China announced its plans to colonize the moon. Technological advances and the discovery of a significant source of water near the lunar poles made this idea even more attractive for commercial structures.

Figure 2. China's Unmanned Lunar Explorer "Chang'e-4" Landed to the Behind of Moon for the First Time in the World on January 3, 2019
Source: http://www.uux.cn/batch.download.php?aid=156511

Chang'e 5 (Chinese: 嫦娥五号) is the fifth lunar exploration mission of the Chinese Lunar Exploration Program, and China's first lunar sample-return mission. Like its predecessors, the spacecraft is named after the Chinese moon goddess Chang'e.

It launched at 20:30 UTC on 23 November 2020 from Wenchang Spacecraft Launch Site on Hainan Island, landed on the Moon on 1 December 2020, collected ~1,731 g (61.1 oz) of lunar samples (including from a core ~1 m deep), and returned to the Earth at 17:59 UTC on 16 December 2020.

Chang'e-5 mission is the first lunar sample-return mission conducted by humanity in over four decades since the Soviet Union's Luna 24 in 1976. By completing the mission, China became the third country to return samples from the Moon after the United States and the Soviet Union.[10]

The Chang'e 5 Ascender lifted off from Oceanus Procellarum at 15:10 UTC, on 3 December 2020, and six minutes later, arrived in lunar orbit.[11]

The Ascender docked with the Orbiter/Returner combination in the lunar orbit on 5 December 2020 at 21:42 UTC, and the samples were transferred to the return capsule at 22:12 UTC. The Ascender separated from the Orbiter/Returner combination on 6 December 2020 at 04:35 UTC.

Figure 3. China's unmanned lunar Explorer "Chang'e-5" landed to the moon on December 1, 2020. After Chang'e 5"lunar Explore collected sample on the moon, it returned to the earth December 16, 2020

Development of a robotic research station near the Moon's south pole.[12]

- Change'6, expected to launch in 2023 or 2024, will investigate the topography, composition and subsurface structure of the landing site, and it will return south polar samples to Earth.[13] Change'7, expected to launch in 2023, is a lander that will explore the south pole for resources.[14] The mission may include an orbiter, a lander and a rover.
- Chane'8, expected to launch in 2027, will verify the utilization and development of natural resources. It may include a lander, a rover, and a flying detector, as well as a 3D-printing experiment using in-situ resource utilization (ISRU) to test-build a structure. It will also transport a small sealed ecosystem experiment. It will test technology necessary to the construction of a lunar science base. It will be automated with robotics to provide the necessary infrastructure and support systems.

Figure 4. China's Space Station Tiangong is Scheduled to Launch in 2022
Source: https://sorabatake.jp/6096

Summarizing the above, it should be emphasized that China's technological developments are taking place at a high speed, the same cannot be said about the development and implementation of national legislation regulating space activities, which is still at the stage of development. Unlike the United States, China lacks a systematic and consistent national space legislation. To date, China has adopted only two low-level regulations on the launch and registration of space objects. With the growth of Chinese space programs, the lack of structured national space legislation may lead to China's inability to fully comply with its international obligations.[15]

7.1.5. China Will Be Explored the Mars, Asteroids, Venus, Jupiter etc. to 2030

China announced that it will launch its first Mars mission probe on July, 2020 year, China Youth Daily reported on January 23 (THU), 2020, adding that this is the first time the country disclosed the launch month of its Mars exploration program. The Mars probe will be sent by the Long March 5 Y4 carrier rocket, said the newspaper, citing sources from the China Aerospace Science and Technology Corporation (中国航天科 技集团有限公司: CASC). The Long March-5 Y4 rocket has recently completed a 100-second test for its high thrust hydrogen-oxygen engine, which is the last engine examination before the final assembly.

According to the CASC, China will send a probe to orbit and land and deploy a rover on Mars. In 2020, the Long March 5 rocket will carry out several missions, including the Mars probe launch and the lunar sample return. A total of 24 high thrust hydrogen-oxygen rocket engine tests will be conducted this year for these missions.[16]

China's space ambitions far exceed any other space faring nation in both range and long-term strategy. This includes an incremental strategy of developing space capacity in sequence.

First, build space capacity for cost effective launch and access.

Second, launch its own permanent space station.

Third, create capacity to dominate lunar space.

Figure 5. China, Deep Space Exploration Roadmap (中国深空探测路线图)
Source: https://www.unoosa.org/documents/pdf/copuos/2018

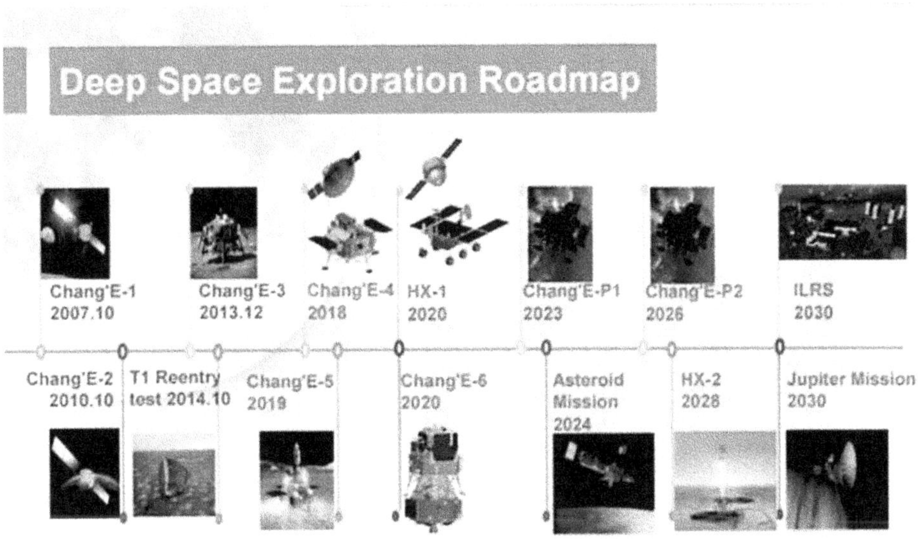

Fourth, once lunar is secured, develop the capacity for sustainable presence on the moon, to include in space manufacturing as well as mature space-based solar power (SBSP) technology to power its lunar base and sustain human presence.

Finally, once that is accomplished, develop capacity for deep space exploration and resource extraction from asteroids.

To meet its extended goals of permanent human presence on the moon and constructing SBSP satellites in LEO by 2025 and in GEO by 2030, China is developing the Long March 9 to be flight tested in 2030.

This rocket development is aimed at that SBSP launch to GEO (by 2030), the launch of the Chang'e 7 (aimed at surveying the Lunar South Pole in 2030), and the Chang'e 8 (2035), that will test key technologies like 3D printing.

According to Li Hong, the Deputy General Manager at China Aerospace Science and Technology Corp (CASC), "the Long March 9 super heavy-lift carrier rocket will be capable of lifting 140 metric tons of payload into a low-Earth orbit, or a 50-ton spacecraft to a lunar transfer orbit.

The giant rocket will also be able to ferry a 44-ton payload to a Mars transfer orbit." CASC estimates that 10 Long March 9 rockets will be required annually to meet China's space goals between 2030 and 2035. The Long March 9 "will be propelled by a new-generation liquid oxygen/kerosene engine with 500 tons of thrust power."

Recently, Chinese scientists discussed and opposed a roadmap and scientific objectives of China's Deep space exploration before space Exploration before 2030.

China's Deep-space Exploration will focus on the key scientific questions of deep space, with an overall planning to explore asteroids, the Sun, Venus, Jupiter *etc.* which starts from Mars.

Figure 6. The Concept Image of China's Mars Probe
Source: https://www.sohu.com/a/325892660_649566

It aims at important research findings in the origin and evolution of the solar system and its planets, disastrous impact on the Earth by the solar activities and small bodies, extra-terrestrial life *etc.*, and promoting astronomy, space science and technology applications. In about 15 years before 2030, 9 or 10 times deep-space exploration missions would be implemented, according to China's existing scientific, aerospace technology and economic basis and development capabilities. All these missions will be executed by three stages.

Figure 7. China, Deep Space Exploration Roadmap
Source: https://www.unoosa.org/documents/pdf/copuos/2018/copuos2018tech19E.pdf

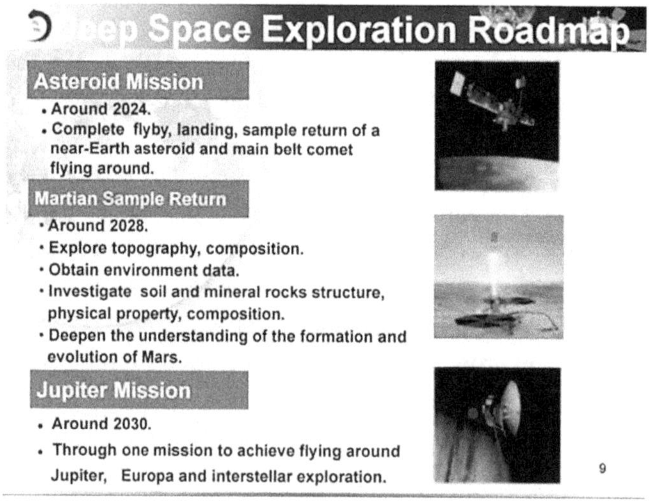

The first stage plans to explore Mars, whose natural environment is close to the Earth's, as well as some asteroids and the Sun which may have disastrous effects on the Earth. This stage includes the Mars Global Remote Sensing orbiter and a small rover, near-Earth asteroid multi-target detection, Sun fixed-point observation. The second solar activities and small bodies, extraterrestrial life *etc.*, and promoting astronomy, space science and technology applications.

The second stage, called the further promotion, will continue to explore Mars, solar activity and main-belt asteroid, hitch includes Venus exploration orbiter, Mars solar polar orbit observations and the main- lander and rover, the belt asteroid (Ceres) sample return. The third stages called sustainable development, on the basis of the first two stages, it includes Jupiter orbiter, Mars sample return and solar storms panoramic observation.

The Chinese space agency is preparing a mid-year mission to Mars, marking the most ambitious project on an exploration checklist intended to achieve equal footing with NASA and transform the nation's technological know-how.

Landing the unmanned probe on the red planet would cap President Xi Jinping's push to make China a superpower in space. The nation already has rovers on the moon, and it's making bold plans to operate an orbiting space station, establish a lunar base and explore asteroids by the 2030s. China's goals in the cosmos include returning samples from Mars in 2028 and exploring Jupiter a few years later, according to a strategy published by the State Council in 2016.

7.1.6. China National Space Administration

The China National Space Administration (CNSA: 国家航天局) is the national agency for China to co-ordinate its space activities. In contrast to most other space agencies worldwide, the organization is not involved with the International Space Station and, in fact, has a small space station of its own.[17]

It is responsible for the national space programme and for planning and development of space activities. CNSA and China Aerospace Corporation (CASC) assumed the authority over space development efforts previously held by the Ministry of Aerospace Industry. This is a subordinate agency of the State Administration for Science, Technology and Industry for National Defense (SASTIND), itself a subordinate agency of the Ministry of Industry and Information Technology (MIIT).[18]The Ninth NPC assigned CNSA as an internal structure of the COSTIND. CNSA assumes the following main responsibilities: signing governmental agreements in the space area on behalf of organizations, inter-governmental scientific and technical exchanges; and also being in charge of the enforcement of national space policies and managing the national space science, technology and industry.

Up to now, China has signed governmental space cooperation agreements with Brazil, Chile, France, Germany, India,[19] Italy, Pakistan, Russia, Ukraine[REMOVED TA FIELD], the United Kingdom, the United States and some other countries. Significant achievements have been scored in the bilateral and multilateral and technology exchanges and cooperation.[20] Administrators of CNSA are appointed by the State Council. CNSA was established as a government institution to develop and fulfill China's due international obligations, with the approval by the Eighth Nation People's Congress of China (NPC).

Figure 8.
Source: http://www.cnsa.gov.cn/english/index.html

7.2. INDIA

7.2.1. The Space Activities Bill

The Space Activities Bill of 2017 has been drafted and is pending approval in the Parliament. This bill would promote, support, and regulate space activities in India by allowing private and non-governmental agencies to involve themselves in space exploration.

Space activities in India, which started in early 1960s, are hitherto pursued by Department of Space (DOS), as normal agency for space activities in India.

As per 'Government of India (Allocation of Business) Rules 1961(as amended from time to time) DOS has been responsible for the space activities in India, for more than five decades, with the major objective of bringing the benefits of space technology and its applications to societal needs and national development.

Pursuance of Space activities were focused on three major areas namely –

1. Space Infrastructure which includes realization of spacecraft for various applications and associated ground infrastructure,
2. Space Transportation systems, which include through realization of various types /class of launch vehicles and associated ground infrastructure including launch facilities, and
3. Space applications for various national requirements through establishment of necessary ground infrastructure and coordination mechanisms. The Department of Space, through Indian Space Research Organization (ISRO) and its R&D centers has been executing various satellite and launch vehicle projects and onceptualized & implemented a number of application programmes for national development and governance.[21]

7.2.2. International Treaties' Obligations on Outer Space Activities

A few space faring nations such as USA, Russia, Ukraine[REMOVED TA FIELD], Republic of Korea and other nations engaged in space activities, such as, South Africa, United Kingdom, Indonesia, Austria, etc. have formulated domestic space legislations. China are in the process of formulating their national space act. France has a Space Authorization Act for providing commercial space activities through Arianespace. It was learnt that the contents, scope and applicability of space legislations of other countries vary a lot, depending upon the nature and Government of India - Department of Space-Draft Space Activities Bill, 2017.

Considering these variations, a Model Law on National Space Legislation was formulated and submitted by International Law Association (ILA), to the UN Committee on Peaceful Uses of Outer Space (UNCOPUOS) in 2013. This Model Law was studied by the Department of Space in India, as it addresses the obligations of a State under UN treaties on outer space activities precisely.

7.2.3. Need for Space Activity Act in India

Based on this reference, the draft Bill on National space activities was taken up for development with necessary customization to match with the Indian context of space activities. On these premises, a draft Space Activities Bill, 2017 has been formulated with due preliminary consultations with a few relevant Ministries. It is intended to have a basic framework with very essential provisions for treaty obligations, with scope for formulating rules and regulations on specific topics / activities.

As the nature of space technologies vis-à-vis scope of space activities change very fast, it would be prudent to revise the rules and formulate with updates to cope with the developments. The Draft Bill includes a provision for Rules Making powers by the Government.

7.2.4. Draft for the Space Activity Bill

The Draft for the Space Activity of India composed of 33 Articles and six Chapters as the following;

Chapter 1, Preliminary Article 1, This Act may be called the Space Activities Act, 2017.
Chapter 2, Space Activity Regulatory Mechanism
Chapter 3, Authorization and License for Commercial Space Activity
Chapter 4, Registration of Space Objects and Liability
Chapter 5, Offences and Penalties
Chapter 6, Miscellaneous

7.2.5. ISRO Sent a Lunar Orbiter, Chandrayaan-1, 2008 & -2, 2019

ISRO sent a lunar orbiter, Chandrayaan-1, on 22 October 2008, which discovered lunar water in the form of ice,[22] and the Mars Orbiter Mission, on 5 November 2013, which entered Mars orbit on 24 September 2014, making India the first nation to succeed on its maiden attempt to Mars, as well as the first space agency in Asia to reach Mars orbit.[23]

On 18 June 2016, ISRO launched twenty satellites in a single vehicle, and on 15 February 2017, ISRO launched one hundred and four satellites in a single rocket (PSLV-C37), a world record.[24] ISRO launched its heaviest rocket, Geosynchronous Satellite Launch Vehicle-Mark III (GSLV-Mk III), on 5 June 2017 and placed a communications satellite GSAT-19 in orbit. With this launch, ISRO became capable of launching 4-ton heavy satellites into GTO.

On 22 July 2019, ISRO launched its second lunar mission Chandrayaan-2 to study the lunar geology and the distribution of lunar water.[25] The Orbiter which was injected into a lunar orbit on 2nd Sept 2019, carries 8 experiments to address many open questions on lunar science. All experiments have been performing well and the data received, suggests excellent capability to deliver on the pre-launch promises. In the period since launch payload teams tuned onboard systems for optimal instrument configurations, derived essential in-flight calibration data, revised / updated data processing steps / software and have started to publish early results. Today, the first set of data are being released for all users.

Figure 9.
Sources: https://www.bbc.com/news/world-asia-india-49589850

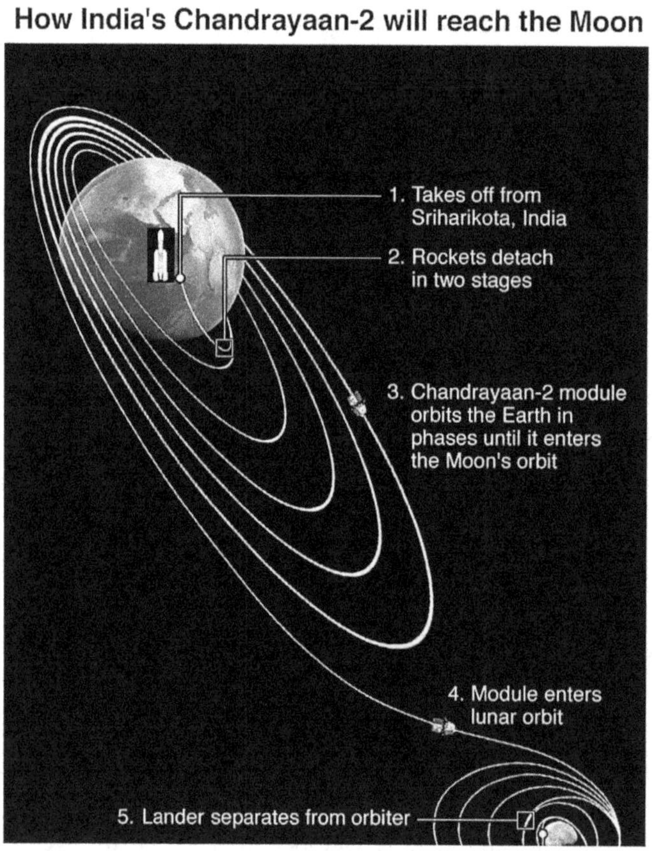

The public release data archived at the Indian Space Science Data Center in Bylalu, near Bengaluru is prepared in the standard, globally followed Planetary Data System 4 (PDS4) format for public release. ISRO Science Data Archive (ISDA) currently holds data sets acquired by Chandrayaan-2 payloads from Sep-2019 to Feb-2020 from seven instruments.[26]

7.2.6. India: Chandrayaan-3, 2021

This will be India's third lunar mission, and its second attempt at landing a spacecraft on the moon. The national agency suffered a setback last year when Chandrayaan-2's lander, Vikram, failed to slow down during its descent and crashed into the luna surface.

In the wake of that failure, the Chandrayaan-3 mission plan will re-attempt a soft landing. Because Vikram (and its rover) were expected to run investigations that would help characterize the water ice reserves at the lunar south pole, it's expected Chandrayaan-3 will carry out these same studies.

Should India stick the landing, it will be just the fourth country in history to put a spacecraft on the surface of the moon.[27]

Marking India's first venture into the interplanetary space, Mars Orbiter Mission (MOM) will explore and observe Mars surface features, morphology, mineralogy and the Martian atmosphere.

Further, a specific search for methane in the Martian atmosphere will provide information about the possibility or the past existence of life on the planet

The enormous distances involved in interplanetary missions present a demanding challenge; developing and mastering the technologies essential for these missions will open endless possibilities for space exploration.

After leaving Earth, the Orbiter will have to endure the Interplanetary space for 300 days before Mars capture. Apart from deep space communications and navigation-guidance-control capabilities, the mission will require autonomy at the spacecraft end to handle contingencies.[28]

7.2.7. ISRO Will Be Explored Mars and Venus

Mars Orbiter Mission 2 (MOM 2) also called *Mangalyaan 2* is India's second interplanetary mission planned for launch to Mars by the Indian Space Research Organisation (ISRO) in the 2021–2022 time frame. It will consist of an orbiter, and may include a lander and a rover.

ISRO plans to return to Mars through this mission. The success of Mangalyaan has prompted the space agency to send a second probe by 2024, which will do deeper studies of the Earth's neighbor and understand the evolution of the red planet better.

The Indian Venusian orbiter mission is a planned orbiter to Venus by the ISRO to study the atmosphere of Venus.

It will be launched some time after 2020. ISRO is planning a mission to the Earth's "twin sister", Venus.

Both the Earth and Venus share similarities in size, mass, density, bulk composition and gravity.

The space agency will fly a spacecraft around 400 km over Venus to conduct research and understand its formation, its atmosphere and its interaction with the solar wind. The mission is expected by 2023.

7.2.8. The Indian Space Research Organization

The Indian Space Research Organization (ISRO) is the space agency of the Government of India and has its headquarters in the city of Bengaluru. Its vision is to "harness space technology for national development while pursuing space science research & planetary exploration".[29]

The Indian National Committee for Space Research (INCOSPAR) was established in the tenure of Jawaharlal Nehru[30] under the Department of Atomic Energy (DAE) in 1962, with the urging of scientist Vikram Sarabhai recognizing the need in space research.

INCOSPAR grew and became ISRO in 1969,[31] also under the DAE.

In 1972, Government of India had setup a Space Commission and the Department of Space (DOS), bringing ISRO under the DOS.

The establishment of ISRO thus institutionalized space research activities in India. It is managed by the DOS, which reports to the prime minister of India.[32]

ISRO built India's first satellite, Aryabhata, which was launched by the Soviet Union on 19 April 1975. It was named after the mathematician Aryabhata.

In 1980, Rohini became the first satellite to be placed in orbit by an Indian-made launch vehicle, SLV-3. ISRO subsequently developed two other rockets: the Polar Satellite Launch Vehicle (PSLV) for launching satellites into polar orbits and the Geosynchronous Satellite Launch Vehicle (GSLV) for placing satellites into geostationary orbits.

These rockets have launched numerous communications satellites and Earth Observation Satellites. Satellite navigation systems like GAGAN and IRNSS have been deployed. In January 2014, ISRO used an indigenous cryogenic engine in a GSLV-D5 launch of the GSAT-14.

Figure 10.
Source: https://gkindiatoday.com/list-of-space-centres-and-units-india/space

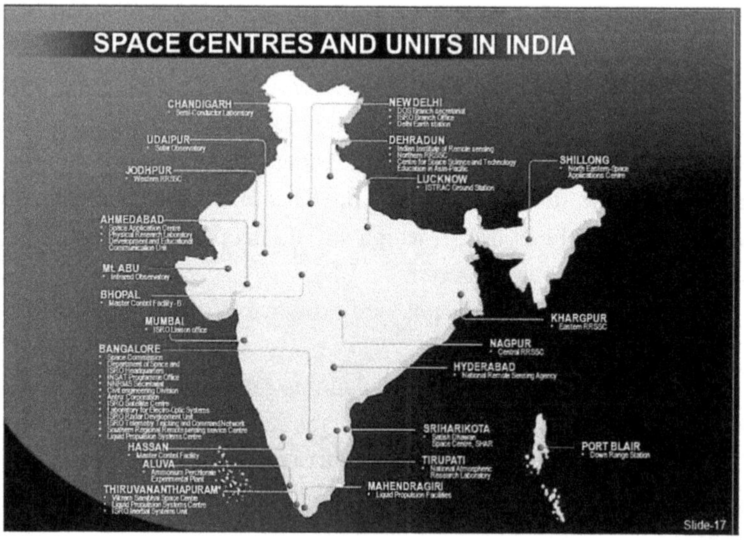

7.3. INDONESIA

7.3.1. Space Activities Act

The Indonesian Space Activities Act of 2,013 (No. 21) composed of 105 Articles, 19 Chapters and it's main key points as the following;

1. Space Activities;
2. Competent Authority;
3. Spaceport Establishment and Its Operation;
4. Safety and Security of Space Activities;
5. Mitigation of Falling Space Objects and Search and Rescue of Astronauts;
6. Registration of Space Objects;
7. International Cooperation;
8. Liability and Indemnification;
9. Insurance, Mortgage and Facility;
10. Environmental Safety;
11. Financing;
12. Society Engagement; and
13. Sanctions.

7.3.2. National Institute of Aeronautics and Space

The National Institute of Aeronautics and Space (Indonesian: *Lembaga Penerbangan dan Antariksa Nasional* or LAPAN) is the Indonesian government space agency.

It was established on November 27, 1963, by former Indonesian president Sukarno after one year's existence of an informal space agency organization.

LAPAN is responsible for long-term civilian and military aerospace research for over two decades, it has managed satellites and domain-developed small scientific- technology satellites Lapan and telecommunication satellites Palapa, which were built by Hughes (now Boeing Satellite Systems) and launched from the US on Delta rockets or from French Guiana using Ariane 4 and Ariane 5 rockets.[33]

It has also developed sounding rockets and has been trying to develop small orbital space launchers. The LAPAN A1 in 2007 and LAPAN A2 satellites were launched by India in 2015.[34]Since 2006 Indonesia and Russia have been discussing the possibility of launching a satellite from Biak island using Air Launch technology.

LAPAN and Russian Federal Space Agency (RKA) have worked on a government-to-government agreement in space cooperation in order to enable such activities in Indonesia. It is planned that Antonov An-124 aircraft will deliver a Polyot space launch vehicle to the new Indonesian spaceport on Biak island (West Papua province). In 2012, discussions resumed. The main stumbling block is Russian concerns over compliance with the terms of the Missile Technology Control Regime; Russia is a co-signatory, Indonesia is not.

7.4. JAPAN

7.4.1. The Basic Space Law(宇宙基本法)

The Japanese Basic Space Law was enacted on May 21, 2008 and then promulgated by the government on May 28, 2008. The Basic Space Law effected into force on August 27, 2008. This Law composed of 35 Articles and five Chapters as the following;

Chapter I, General Provisions (總則, Article 1~12)
Article 1, Purpose (目的)
Article 2, Peaceful Use of Outer Space
Article 6, International Cooperation (国際協力)
Article 8, Responsibilities of State
Chapter 2, Basic Measures (Article 13~23)
Article 13 (Use of Satellites to Contribute to the Improvement of the Lives of the Citizenry)
Article 15 (Independent Launching of Artificial Satellites, etc.)
Article 20 (Preservation of the Environment)
Chapter 3, Basic Space Plan (宇宙基本計画, Article 24)
Chapter 4, Strategic Headquarter for Space Development (宇宙開発戦略)
Article 25~34)
Article 25 (Establishment)
Article 27 (Organization, 組織)
Article 28 (Director-General of the Headquarters for Space Development)
Article 30 (Members of the Headquarters for Space Development)
Article 31 (Submission of Materials and Other Forms of Cooperation)
Article 32 (Affairs)
Chapter 5, (Enactment of Legislation with regard to Space Activities)

On May 21, 2008, the Basic Space Law (Basic Law) was enacted when it was passed by the Lower House of the Diet following its earlier ratification by the Upper House.

This law is Japan's first basic law relating to space activities since 1970 when Japan initiated its space development with the launch of Ohsumi, a satellite for testing technology.

The Basic Space Law consists of the main body comprised of a total of 35 articles and four supplementary provisions. The fundamental principles of the legislation are detailed below.

The objectives of the law are to improve the lives of the people, develop the economy and society, and contribute to world peace and human welfare by establishing the basis of Japan's space development and use of space as well as its responsibility in these areas and by formulating a basic plan (Article 1).

Furthermore, Japan will engage in space activities that will contribute to the formation of a safe and secure society Japan, the peace and security of international society, and the security of Japan (Article 3).

"The Basic Plan on Space Policy" is formulated to propel policies regarding Japan's space development and use, comprehensibly and systematically based on Article 24 of the Basic Space Law (enacted in 2008, Law No. 43), and is considered to be the most fundamental plan of space exploitation. National Space Policy Secretariat plans and designs policies to be incorporated in the Basic Plan, including those discussed in the Committee on National Space Policy.[35]

The Space Development Strategy Headquarters (宇宙開発戦略本部) was established in the Cabinet to formulate the basic policies for promoting space development and the measures to be implemented comprehensively and systematically in space development as the Basic Space Plan (Article 24-25).

The head of the Space Development Strategy Headquarters is the Prime Minister, and the Deputy Head will be the Secretary-General of the Cabinet Secretariat and the Minister of Space Development (Article 28-29).

After the enactment of Basic Space Law in 2008, the Strategic Headquarters for Space Development (SHSD) was set up in the Cabinet, led by the Prime Minister himself.

As for the secretariat, following several structural changes made within the government since then, the National Space Policy Secretariat (NSPS) set up in Cabinet Office is currently serving SHSD since 2016. After its establishment, SHSD worked out a *Basic*

Plan on Space Policy in 2013 (and revised it in 2016), which is considered the masterplan for space activities by the Japanese government.

They also began publishing the *Implementation Plan of the Basic Plan on Space Policy* annually, breaking down policy agendas considered in the Basic Plan into concrete tasks and showing precise time schedules to achieve them.

The tasks described in the Implementation Plan are not limited to domestic policies but also include some international aspects, such as plans to further enhance space cooperation frameworks, especially in the Asia-Pacific regions.[36]

7.4.2. The Space Activity Act

Two outer space-related laws were promulgated in Japan's official gazette on November 16, 2016: the Act on Launching Artificial Satellites and Managing Satellites (人工衛星等の打ち上げ及び衛星管理に関する法律:Space Activity Act, Act No. 76 of 2016) and the Act on Securing Proper Handling of Satellite Remote Sensing Records衛星リモートセンシング記録の適正な取扱いの確保に関する法律:Remote Sensing Records Act, Act No.77 of 2016). (Texts of both laws, KANPO 官報), extra edition No. 252 (Nov. 16, 2016), at 3, KANPO website in Japanese.

Space Activity Act composed of sixty five Act and eight Chapters.

Most of the provisions of these laws will become effective within two years of the promulgation date. (Space Activity Act, Supp. Provisions, Article 1; Remote Sensing Records Act, Supp. Provisions, Article 1.)

The Space Activities Act concerned with the launch and management of satellites, and it consists of 3 parts –① launch approval for satellites, ② management approval for satellites, and most importantly, ③ third party liability in the case of a launch failure or a de-orbit of a satellite.

One of the biggest obstacles in promoting commercial space activities is the launch liability issue, where the cost of a launch failure poses a great risk for any private entity to undertake. With the advent of new, commercial space enterprises such as *SpaceX*, the Japanese government hopes to alleviate the financial concerns and make it easier for companies to launch satellites and participate in various space activities.

The Space Activity Act is linked to Articles 6 and 7 of the Outer Space Treaty, which have to do with international responsibility for national activities in outer space and liability for damage.

In creating the new Act, the Japanese government referred to the systems adopted by other developed countries regarding third party compensation for damages by satellites and the management of satellites launched by private parties.[37]

The Satellite Act states that a person who launches a satellite from the territory of Japan, or from a ship or airplane registered in Japan, must obtain permission from the Prime Minister prior to the launch (Space Activity Act, Article 4).

To obtain the permission, the person must have a certificate for a rocket design and for radio equipment at a launching facility.

In addition, the ability to launch a rocket safely and the purpose for the satellite launch must be evaluated. (Article. 6) Managing a satellite from Japan also requires permission from the government (Article 20).

A person who launches a rocket must have insurance for any potential damage arising from accidents, and the government is to supplement the potential compensation to allow for damage that cannot be covered by private insurance. (Article 9 and 40.)

As noted above, any person who plans to launch a satellite from a launching facility located in the territory of Japan must obtain permission for each launch from the Prime Minister of Japan. No permission is required for any launch from outside the territory of Japan (which may instead be subject to the licensing regime of the appropriate country).[38]

The Space Activities Act established special civil liability rules for third-party damages caused by space activities. In establishing those rules, the lawmakers considered the necessity of protecting the public from inherently dangerous space activities, the extreme difficulty of victims being able to prove the negligence of a space operator that caused harm, and the principles applied to international liabilities under the Convention on International Liability for Damage Caused by Space Objects.

Japan's Space Activities Act provides that the launch operator bears liability for accident damages even if they are due to problems in the payload.

This channeling of liability would seem to be disadvantageous to launch operators, but it can be expected to enhance the competitive position of the Japanese companies providing this service, because it reassures customers around the world who are seeking to have their satellites put into orbit. France is the only other country that has adopted a similar provision.

7.4.3. The Remote Sensing Records Act

The Remote Sensing Records Act composed of thirty eight Articles and seven Chapters. The Remote Sensing Records Act establishes regulations for satellite remote sensing data. A person who plans to use a satellite remote sensing device through radio equipment located in Japan must obtain permission for each device from the Prime Minister (Remote Sensing Records Act, Article 4.) The user must implement measures to prevent unauthorized use of the equipment. (Article 8) The user must also stop remote sensing when the satellite is out of the orbit for which permission was granted to collect information (Article 9.)

Reception of remote sensing data must be carried out only by the facility listed on the permit (Article 10.) Satellite remote sensing data can be provided only to the person who obtained the license to operate the remote sensing device from the government, except in cases of an emergency. (Article 18 and 21.)

The Prime Minister can prohibit transfer of particular remote sensing data for a limited time if it is necessary for national security (Article 19).

The Prime Minister supervises persons who use satellite remote sensing devices and who hold satellite remote sensing data, and the Prime Minister can conduct on-site inspections and issue orders with respect to the operation of the devices and the data (Article 27 and 29.).[39]

The Satellite Remote-Sensing Act is pertained with the commercial use of sub-meter satellite images, and the aim is to put regulations in place to prevent images being handed to countries with malicious intent or terrorist groups.

The proposed act stipulates an approval process for the use of remote-sensing instruments and acquiring images, and also gives the government the right to restrict the distribution of the images to foreign entities. Thus, in order to foster the growth of the EO data industry in Japan, the Japanese government saw the need to follow the footsteps of Europe and the US to establish clear guidelines on the sales and distribution of satellite observation data.

7.4.4. Japanese Lunar Mission: Hakuto-R Mission 1 and 2, 2021 and 2023

Japanese company ispace is no stranger to big ideas (it once suggested putting a giant billboard on the surface of the moon), but it hasn't had much success in actually getting off the ground. Its lunar program, Hakuto(ハクト)-R, is now being run in a partnership with Draper Labs. two companies plan to launch a lunar lander in October 2021, and a lander and a rover combo in March 2023.[40] Both missions will seek to validate the landing technologies of the spacecraft, and scout for potential water ice and key resources that could be mined. HAKUTO-R is the world's first commercial lunar exploration program.

Figure 11. Landing to the Moon by Hakuto-R Mission 1 and 2, 2021 and 2023
Source: https://ispace-inc.com/hakuto-r

It includes ispace's first two lunar missions: Mission 1, a soft lunar landing in 2021, and Mission 2, a lunar landing and deployment of a rover for lunar surface exploration 2023. For both missions, ispace's lander will be a secondary payload on SpaceX's Falcon 9 rocket.

7.4.5. Japan: Smart Lander for Investigation Moon, January 2022

This will be Japan's first mission to the lunar surface. If JAXA is keen to play an important role in the moon rush of the decade, this mission is critical in demonstrating the agency has what it takes to actually land safely. SLIM (Smart Lander for Investigating Moon) will use the same technologies that power facial recognition systems to recognize lunar craters and determine its location as it attempts to touch down near the Marius Hills Hole. Thought to be an ancient lava tube that's almost 300 feet deep, the hole could provide protection from radiation for a future human outpost. A lunar rover with two inflatable wheels might hitch a ride on this mission and explore this mysterious tunnel.[41]

An illustration of Japan's Smart Lander for Investigating the Moon, or SLIM. The moon lander is scheduled to launch in fiscal 2021.

Figure 12.
Source: Collaboration-needed-future-space-missions/#.XrOpwIt7kfh

7.4.6. Japan Eyes Using Moon Water as Fuel for Space Exploration in 2030s

Japan's space agency announced that it aims to engage from around the mid-2030s in lunar exploration using hydrogen generated by water extracted from the Moon's ice posits as fuel on September 28, 2020.[42] Using the water-derived fuel is expected to cut costs compared with transporting fuel from Earth. While liquid water does not exist on the Moon's surface, past research suggests there may be ice at a crater near the lunar south pole, which has never been exposed to sunlight. Japan plans to work with the United States in building the lunar orbit space station Gateway in the 2020s and construct by around 2035 a fuel factory at the lunar south pole.

The fuel will be used in a reusable spacecraft able to carry four astronauts to and from the Gateway and a transport vehicle that can travel up to 1,000 kilometers on the Moon, according to the Japan Aerospace Exploration Agency.

Water-derived fuel is created by first splitting water into its components of oxygen and hydrogen using a solar cell. Energy is created by recombining them.

This Supplied CG image shows a concept for a base on the Moon

Figure 13. Supplied computer graphic shows image of a base on the Moon[43]

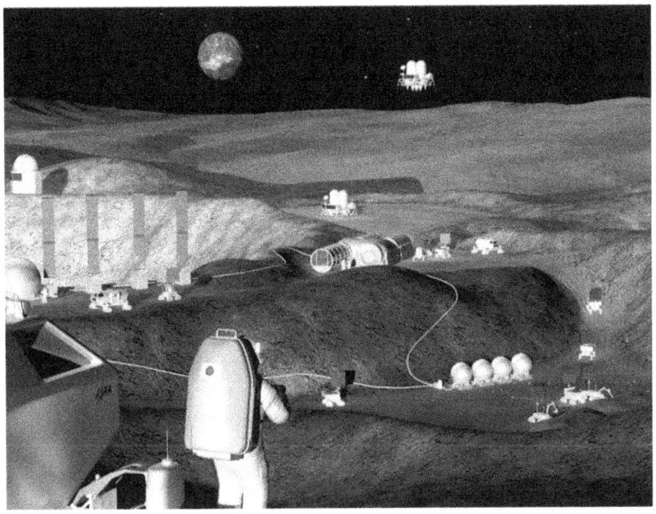

According to JAXA's estimates, 37 tons of water is necessary for a trip to and from the Gateway, while 21 tons is needed for each surface exploration trip. The agency believes it would be cheaper to use fuel created with the Moon's water than bringing water from Earth if five to seven manned exploration trips are made.

Other countries, including India and the United States, are intent on analyzing the water resources on the Moon. China, which has already landed an unmanned spacecraft on the lunar surface, plans to send a probe later this year to the Moon to collect soil samples.

7.4.7. JAXA's Mars Exploration

The Martian Moons Exploration (MMX) is a robotic space probe set for launch in 2024 to bring back the first samples from Mars' largest moon Phobos.[44] Developed by the Japanese Aerospace Exploration Agency (JAXA) and announced on 9 June 2015, MMX will land and collect samples from Phobos once or twice, along with conducting Deimos flyby observations and monitoring Mars' climate.

The mission aims to provide key information to help determine whether the Martian moons are captured asteroids or the result of a larger body hitting Mars. The Japan Aerospace Exploration Agency and other Japanese government officials officially approved the MMX project to proceed into development on 19 February 2020, according to a post on JAXA's website.[45]

Phobos is the largest moon of Mars. The spacecraft will enter orbit around Mars, then transfer to Phobos, and land once or twice and gather sand-like regolith particles using a simple pneumatic system. The lander mission aims to retrieve a minimum 10 g (0.35 oz.) of samples. The spacecraft will then take off from Phobos and make several flybys of the smaller moon Deimos before sending the Sample Return Capsule back to Earth, arriving in July 2029.[46]

The mission architecture uses three modules: propulsion module (1800 kg), exploration module (150 kg) and the return module (1050 kg).

With the mass of Deimos and Phobos being too small to capture a satellite, it is not possible to orbit the Martian moons in the usual sense.

However, orbits of a special kind, referred to as quasi-satellite orbits, can be sufficiently stable to allow many months of operations in the vicinity of the moon.

Development and testing of key components, including the sampler, is ongoing. As of 2020, MMX is scheduled to be launched in September 2024, and will return to Earth five years later. For sample collection, the mission opted to use an air gun to puff pressurized gas, pushing about 10 grams of soil into the sample container.

7.4.8. JAXA Will Be Explored Venus, Asteroids, Mercury, Jupiter, Saturn etc

7.4.8.1. JAXA's Venus Climate Orbiter Akatsuki

The Venus Climate Orbiter Akatsuki is a spacecraft developed in order to unravel the mysteries of the atmosphere on Venus. The Venus spacecraft Akatsuki (金星探 査機 「あかつき」)was launched from the Tanegashima Space Center (種子島宇宙 センター) on May 20, 2010 (UTC) and became the first Japanese probe to enter orbit around a planet other than the Earth on December 7, 2015.

Akatsuki is equipped with five cameras capable of obtaining images of Venusian atmosphere in the infrared, visible, and ultraviolet ranges, and also with an ultra-stable oscillator used to measure the vertical profiles of temperature and other factors. It traces an elliptical orbit around Venus in about 10 days and allows for investigating the flow and constituents of the atmosphere, as well as the presence of lightning and volcanic activity.

Akatsuki's target of Venus is a planet approximately the same size as our own and considered the "sibling" to the Earth. However, its atmosphere is about 100 times as thick as that of the Earth's, and it is covered in high-temperature carbon dioxide. The "super rotation" reaching 100 m/s at high altitudes is a fierce wind the cause of which remains unknown.

Akatsuki takes advantages of the six instruments to observe meteorological phenomena on Venus in detail. These findings are expected to lead to a better understanding of meteorological phenomena not only on Venus, but also on other planets and further help explicate why the Earth's atmosphere is the way it is, as well as how it may change in the future. [47]

7.4.8.2. JAXA's Hayabusa 2 Touched Down on the Ryugu Asteroid

Japan is launching multiple missions to explore the mysteries of the solar system in the coming years, joining hands with the European Union and countries such as India to compete with space superpowers such as the United States and Russia. The ultimate goal of space exploration is "to expand the areas of activities for humans and find another habitable planet. In 2018, The mission will return the samples to Earth in late 2020.[48]

The probe's aim is to collect a sample of sand on the asteroid that scientists believe could include some of the "raw materials" of the stars. This is based on the fact there is no atmosphere on the asteroid so its surface is thought to have remained almost unchanged since the birth of the solar system. Though contributions made by Hayabusa 2 and other missions last year have given momentum to Japan's space

Figure 14. White Colure Stars are many Asteroids Belt.
Source: https://sciencesprings.wordpress.com/tag/nasa-lucy

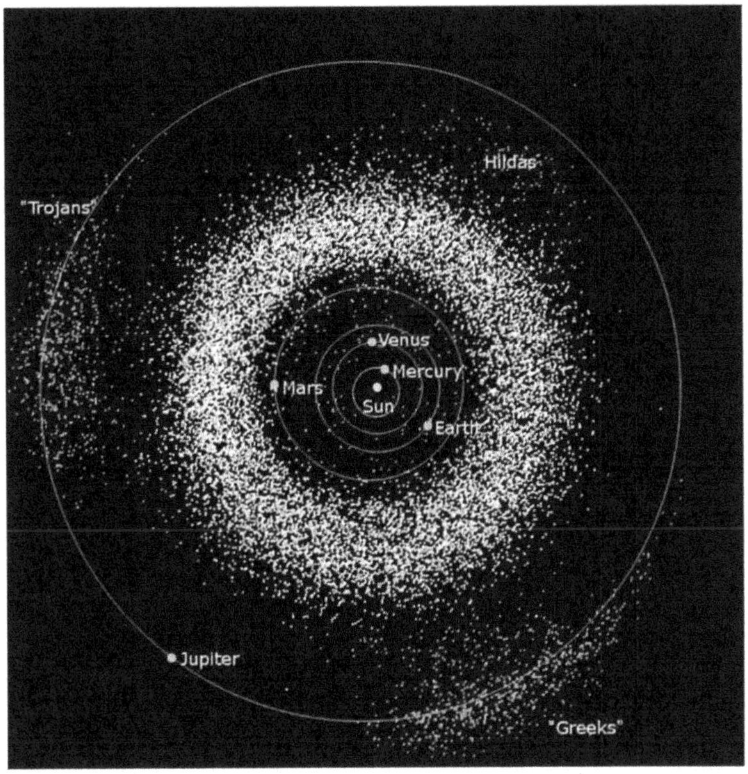

exploration, the country went through a series of bitter failures in the late 1990s and early 2000s. In 2010, Hayabusa 2's predecessor, Hayabusa, returned to Earth with surface samples from an asteroid for the first time ever. This paved the way for more missions. They are found in major asteroid belts and orbit between Mars and Jupiter.

Space missions require highly advanced technologies and a massive amount of funds, creating the need for more international collaboration

7.4.8.3. JAXA's Mercury Project

In October on 2019, the Japanese space agency worked with its European counterpart and successfully launched a spacecraft as part of efforts to explore Mercury. The probe is now on a 9 billion-kilometer voyage in space that will take seven years to complete. The Mercury project is aimed at observing the planet's atmosphere, magnetic field and surface to shed light on how the planet was formed.

In fiscal 2021, Japan plans to launch DESTINY+ (Destiny Plus) to observe dust in space that delivered organisms to Earth. JUICE, a craft that will orbit Jupiter and its moons, and MMX, designed to bring back materials from a Mars moon, are slated for launch in fiscal 2022 and 2024, respectively.[49]

Japan has been exploring ways to work together with different partners, rather than just following the United States. Johann-Dietrich Worner, director general of the European Space Agency (ESA) welcomed the collaboration with Japan on exploring Mercury, saying it is very valuable.

Japan has also decided to work with India to study ice on the moon, said Hiroshi Sasaki, head of the JAXA Space Exploration Center. JAXA selected MIO as new name for the Mercury Magnetosphere Orbiter (MMO: 水星磁気圏探査機) to be launched on 2018 Japanese fiscal year.

Selection process is based on public response to the MMO renaming project made during the designated correspondence period, February 20 to April 9, 2018.[50]The Mercury Magnetosphere Orbiter (MIO, a probe developed in a joint project between JAXA and ESA, will observe the weak magnetic field that covers Mercury. The idea is to understand how strong a magnetic field needs to be for life to exist by comparing Mercury and Earth. However, due to the difficulty of entering orbit and the intensely hot environment created by the strong sunlight that is ten times as strong as it is on Earth, no probe has been sent to Mercury in close to 30 years, when the Mariner 10 was sent.

BepiColombo (国際水星探査計画), an international mission for the study of Mercury, is the first large scale cooperative project between Japan and Europe that will make comprehensive and multifaceted observations of the magnetic field, magnetosphere, internal structure, and surface of Mercury for the first time.

Mercury and the Earth are the only planets with their own magnetic fields among the terrestrial planets, which also include Mars and Venus. In comparing these two planets, it may be possible to obtain clues to understanding Earth's magnetic field and magnetosphere as well as the various magnetospheres which exist in space. Mercury also has a special structure not seen in other planets such as an enormous core that is three quarters of the radius. .[51]

7.4.8.4. JAXA's Jupiter's Project

The Jupiter Magnetospheric Orbiter (JMO: 木星磁気圏オービター) is a cancelled space probe proposed by the Japanese Aerospace Exploration Agency (JAXA), to undertake detailed *in situ* studies of the magnetosphere of Jupiter as a template for an astrophysical magnetized disk. In November 2013, Japan's JAXA scientists stated that ″it is difficult for JAXA to launch the outer planet mission by itself to meet the time constraint for cooperative observation with JUICE.

Therefore, direct collaboration with European groups/team should be necessary for Japanese scientists to participate in this attractive Jupiter system mission.″ The outcome was the selection of four Japanese instrument development teams for 4 of the 10 "Jupiter Icy Moon Explorer (JUICE) mission" instruments.[52]

The JUICE spacecraft is set for launch in June 2022 and will reach Jupiter in October 2029 after five gravity assists and 88 months of travel.[53]

7.4.8.5. JAXA's Saturn Project

Like at other planetary magnetopauses, the solar wind that encounters Saturn's magnetopause has been processed by the planetary bow shock, making it hotter and denser. However, compared to conditions at Earth's magnetopause, some important (dimensionless) plasma parameters are significantly different at Saturn's magnetopause, In the 2030s, one of Japan's goals in space is to study Saturn, said Kuninaka Hitoshi Kuninaka, a vice president of the Japan Aerospace Exploration Agency (JAXA). "It takes longer to get there, while there is less sunlight to generate electricity for communication and observation. So we need to establish an energy-saving technology, which I believe will also help our lives on Earth," he said.[54] because Saturn's bow shock is much stronger than Earth's, as we have already discussed.[55]

Figure 15.
*Source: https://www.essc.esf.org/fileadmin/user_upload/essc/Toukaku_JAXA_*

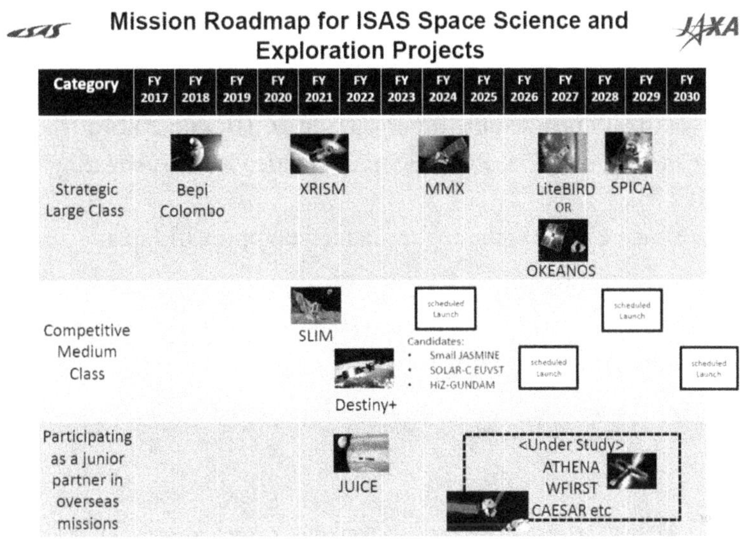

In summary, our understanding of space plasmas in the Solar System is constantly evolving. Thanks to spacecraft like Cassini that are exploring planetary systems in detail for the first time, we can make significant fundamental plasma physics discoveries.

As well as being highly relevant for research into solar system space environments, the current wealth of in situ measurements made by Cassini (Saturn Spacecraft) and other spacecraft represent a valuable source of information about distant astrophysical environments that are of great scientific interest.

The international Cassini mission has concluded its remarkable exploration of the Saturnine system in spectacular style, by plunging into the gas planet's atmosphere.[56]

7.4.9. JAXA

The Japan Aerospace Exploration Agency (JAXA: 国立研究開発法人宇宙航空 研究開発機構, literally "National Research and Development Agency on Aerospace Research and Development") is the Japanese national aerospace and space agency.

The JAXA was based on the "*Law Concerning Japan Aerospace Exploration Agency*" (Law Number 161) on December 13, 2012.

Through the merger of three previously independent organizations, JAXA was formed on 1 October 2003. JAXA is responsible for research, technology development and launch of satellites into orbit, and is involved in many more advanced missions such as asteroid exploration and possible human exploration of the Moon.[57]

Its motto is *One JAXA*[1] and its corporate slogan is *Explore to Realize* (formerly *Reaching for the skies, exploring space*).[58]

The Basic Space Law was passed in 2008, and the jurisdictional authority of JAXA moved from MEXT to the Strategic Headquarters for Space Development (SHSD) in the, led by the Prime Minister.

In 2016, the National Space Policy Secretariat (NSPS) was set up the Cabinet. On 14 September 2007, JAXA succeeded in launching the lunar orbit explorer Kaguya, also known as SELENE (costing 55 billion yen including launch vehicle), the largest such mission since the Apollo program, on an H-2A rocket.

Its mission is to gather data on the Moon's origin and evolution. It entered lunar orbit on 4 October 2007.

After 1 year and 8 months it impacted the lunar surface on 10 June 2009 at 18:25 UTC. JAXA plans to launch its first lunar surface mission, SLIM (Smart Lander for Investigating Moon) on an Epsilon rocket in fiscal year 2019.[59]

JAXA Tanegashima Space Center is the rocket-launch complex in Japan.

Figure 16.
Source: https://global.jaxa.jp/about/centers/tnsc/index.html

On 7 December 2010, Akatsuki was unable to complete its Venus orbit insertion maneuver. Akatsuki finally entered Venus orbit in 7 December 2015, making it the first Japanese spacecraft to orbit another planet, sixteen years after the originally planned orbital insertion of Nozomi.

Japan has plans to land astronauts on the moon by 2030- with a little help from the United States.

The Japanese space agency JAXA said it envisions human missions to the moon, potentially to study and make use of water ice deposits at the lunar poles. The JAXA plan, though, would involve making use of NASA's proposed Deep Space Gateway in lunar space, which would serve as the jumping-off point for expeditions to the lunar surface.

7.5. PHILIPPINE

7.5.1. Philippine Space Act

An Act Establishing the Philippine Space Development and Utilization Policy and Creating The Philippine Space Agency, and for Other Purposes (Space Act: Republic Act No. 11363)[60] was signed by the President of Philippine government on August 8, 2019. This Act composed of twenty nine Sections such as main key point of it as the following;

Section 1. Short Title. -This Act shall be known as the "Philippine Space Act".
Section 5. Framework of the Space Policy
Section 8. Powers and Functions.
Section 10. Director General of the PhilSA.
Section 13. Philippine Space Council.
Section 14. Transfer of Astronomical and Space-related Functions
Section 19. Annual Report.
Section 22. Philippine Space Development Fund
Section 23. National Registry of Space Objects
Section 24 . Liability of the Philippines for National Space Objects.
Section 29. Effectivity.

This Act shall take effect fifteen (15) days after its publication in the *Official Gazette* or in any two (2) newspapers of general circulation in the Philippines

7.5.2. Philippine Space Council

A Philippine Space Council, to be chaired by the President and composed of other key members of the Cabinet shall be created to act as the principal advisory body to the Director General. It is proposed that at least 30 hectares (74 acres) will be allocated as PhilSA's headquarters and main research facility in an official site within the Clark Special Economic Zone in Pampanga and Tarlac with a target completion date of 2022. Additional facilities (launch sites, research and development) are also to be established in the future.[61]

The Philippine Space Council will be the primary advisory body.

It is to be headed a chairperson, a position to be held by the President of the Philippines.

The following will be the composition of the space council:[62]

- Chairperson – President of the Philippines
- Vice Chairpersons – Secretaries of Science and Technology and National
- Defense Other members
- Senate Committee Chair on Science and Technology
- House Committee Chair on Science and Technology
- Director General of the National Economic and Development Authority

7.5.3. Philippine Space Agency

The Philippine Space Agency, or PhilSA, is the national space agency of the Philippines. The unified space agency is defined by the "Philippine Space Act" (*Republic Act No. 11363*) which was signed into law on 8 August 2019, intended to manage and operate the Philippine government's space program which was decentralized and handled by various agencies of the Department of Science and Technology (DOST).

According to the Department of Science and Technology, the Philippines already possess enough infrastructure to run a dedicated space agency.[63]

Since 2010, it has spent ₱7.48 billion (or $144 million) for space research and development, aided 5,500 scholars, trained more than 1,000 space science experts, and established 25 facilities in various parts of the Philippines.

It has also developed three small-scale satellites (Diwata-1, Diwata-2, and Maya-1) and had them launched and deployed to space.

While the immediate goals of the agency will not involve launching its own rockets like NASA and JAXA, it is planned that the agency would pursue such goals in the long term.[64]

7.6. THE DEMOCRATIC PEOPLES REPUBLIC OF KOREA (NORTH KOREA)

7.6.1. The National Aerospace Development Administration

The North Korean Committee of Space Technology (北鮮宇宙空間技術委員會) was the agency of the government of the Democratic People's Republic of Korea (North Korea) responsible for the country's space program. The agency was terminated and succeeded by the National Aerospace Development Administration (NADA: 國家宇宙開發局) on April, in 2013 after the Law on Space Development was passed in the 7th session of the 12th Supreme People's Assembly. [65]

The Act sets out the North Korean principles of peaceful development of space determines compliance with the principles of the Juche ideology (주체사상, 主体思想, North Korean ideology) and independence, as well as the aim of solving scientific and technological problems of space to improve the economy, science and technology.

The law also regulates the position of the NADA and the principles of notification, security, research and possibly compensation in relation to satellite launches.

The law calls for cooperation with international organizations and other countries, the principle of equality and mutual benefit, respect for international law and international regulations for space. The law also opposes the militarization of space. North Korea entered to the Outer Space Treaty of 1967 and Registration Convention of 1975 on March 12, 2009 and the after a previous declaration of preparations for a new satellite launch.[66]

The KCST operated the Tonghae(동해, 東海) Satellite Launching Ground and Sohae (서해, 西海) Satellite Launching Station rocket launching sites, Baekdusan (백두산, 白頭山)-1 and Unha (은하, 銀河) (Baekdusan-2) launchers, Kwangmyŏngsŏng (광명성, 光明星) satellites.

7.6.2. North Korean Missile Test

There have been a number of North Korean missile tests. North Korea has also fired a number of short-range missiles into the Sea of East (Sea of Japan), in what have been interpreted as political gestures.[67] As of 20 March 2020, North Korea has carried out 147 tests of strategic missiles since its first such test in 1984.[68] Under Kim Jong -un, 119 tests have been undertaken as of December, 2019.[69]

Taepodong-2 (TD-2, Korean language, 대포동 2호, 大浦洞2号) is a designation used to indicate what was initially believed to be a north Korean two or three-stage ballistic missile design that is the successor to the Taepodong-1 technology demonstrator.

Figure 17.

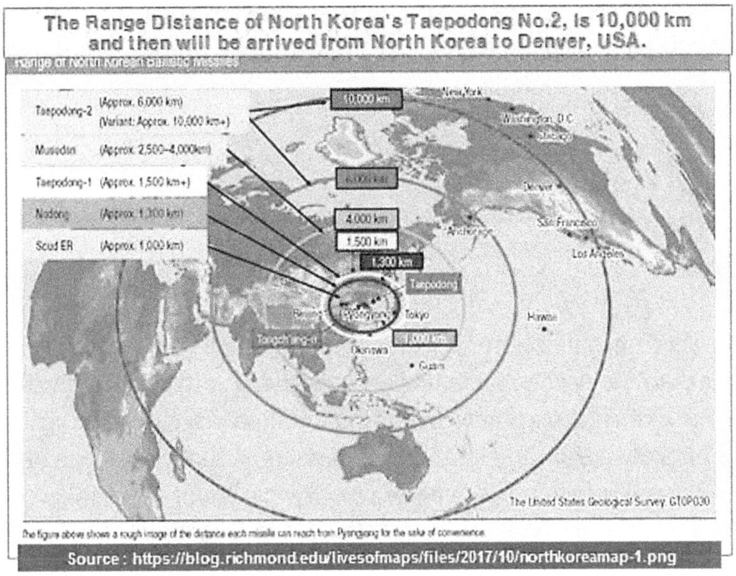

Based on the size of the missile, the fuel composition, and the likely fuel capacity, it is estimated that a two-stage variant would have a range of around 4,000 km (2,500 statute miles) and a three-stage variant would be capable of reaching as far as 4,500 km (2,800 statute miles), giving it potentially the longest range in the North Korean missile arsenal.

The burn time of each stage is a little over 100 seconds, thus allowing the missile to burn for 5 or 6 minutes. Speculative variants of the missile could be capable of a range distance of 10,000 km.[70]

7.6.3. North Korean Lunar Exploration Programme

The first phase of the NKLEP. South Korean internet paper Jaju Minbo(자주민보: 自主民報) carried an article on March 23, 2012, that North Korea is likely to launch a lunar explorer satellite.[71]

The second phase of the NKLEP after orbiting. In an interview with the Associated Press, a senior official said on Aug. 4, 2016, that the North Koreans will start designing lunar orbiter and lunar lander immediately after launching a GEO communication satellite as planned in the Second Space Development Five-Year Plan.

The development and completion of the Paektusan-1 rocket engine was an epoch-making event that advanced the moon exploration plan. North Korea stated its ambition to land a probe on the Moon.

Prerequisite paving the way for any manned moon landing, the third phase of the NKLEP after orbiting and landing would be to return rock samples to the Earth with a robotic probe.

Already illustrated back in 2015's Pyongyang Science-Technology Complex, by a schematics of the orbital trajectory of the North Korean Lunar Exploration Program Phase III, the lunar sample-return mission. Launched with a heavy Unha-20 booster able to place 5t into a LTO, sometimes after 2026. Splash landing of the return capsule in the Pacific Ocean.

Mr. Hyon Kwang Il, director of the scientific research department of North Korea's National Aerospace Development Administration, said that North Korea also intends "to do manned spaceflight and scientific experiments in space, make a flight to the moon and moon exploration and also exploration to other planets.[72]

7.7. THE REPUBLIC OF KOREA

7.7.1. Introduction

The space industry has become an important sector in Asia, which will develop with greater rapidity in the 21st century. Thanks to the development and application of artificial intelligence, mobile communication industry, increase of space exploration and exploitation activities, the space industry will grow continuously and the market size of the global space industry is likely to see an annual.

The growth rate of the space industry is expected to grow by about 10% globally, making it a popular industry in the future. For this reason the Korean Government decided to actively foster the aerospace industry.

The growth rate of the space industry is expected to grow by about 10% globally, making it a popular industry in the future. For this reason the Korean Government decided to actively foster the aerospace industry.

The Korean space policy is based on the national space program as well as the space relating law in Korea which is divided into three branches as follows;

1. Aerospace Industry Development Promotion Act of 1987,
2. Space Development Promotion Act of 2005,
3. Space Damage Compensation Act of 2007.

It deals with the development and the promotion of the aerospace industry and pursues the objective of controlling accident mitigation and the regulation of compensation in case of damage. I introduce the legislative history and contents main points of the Korean three space acts as the following.

7.7.2. The Aerospace Industry Development Promotion Act

The Aerospace Industry Development Promotion Act (항공우주산업개발촉진법: 航空宇宙産業開促進法) was passed by the majority of the Korean National Assembly and was proclaimed with Law No. 8852 by the Government on December 4, 1987. [73]This abovementioned Act has been amended 22 times during the more 33 years until on March 30, 2020 and it composed of 22 Articles.

It is the purpose of this Act to contribute to the sound development of the national economy and the improvement of national life by rationally supporting and promoting the aerospace industry as well as research and effective development with regard to aerospace science and technology. In this framework the Korean Ministry of Trade, Industry and Energy has always been an important player and handled kindly the budgetary matters and financial support for manufacturing of aircraft, satellites and spacecraft.[74]

The Aerospace Industry Development Promotion Act was composed of as the following main items.

1. Purpose (Art.1)
2. Definition (Art.2)
3. Establishment of the Master Plan for Aerospace Industry Development (Art.
4. Fosterage of Aerospace Industry (Art. 4).
5. Designation of and Support for Specialized Complex for Aerospace Industry (Art.8-2).
6. Performance and Quality Inspection (Art. 10).
7. Restriction etc. on Use (Art. 11).
8. Financial Support (Art.12).
9. Lending etc. of State-Owned Facilities, Apparatus etc. (Art. 13).
10. Establishment of Aerospace Industry Development Policy Council (Art. 14).
11. Composition, etc. of Council (Art. 16).
12. Delegation and Entrustment of Authority (Art. 19).
13. Inspection Fee (Art. 20).
14. Penalty Provisions (Art. 21).
15. Joint Penalty Provisions (Art. 22).

7.7.3. The Space Development Promotion Act

First of all I had proposed strongly to the Ministry of Science and Technology of the Korean government and Korea Aerospace Research Institute (KARI) that we must legislate a "*Draft for the Space Development Promotion Act*" in order to develop efficiently and reasonably the Korean space industry in 2003.[75]

The Korean Government was enacted a new "Space Development Promotion Act(우주개발진흥법: 宇宙開發振興法)" on May 2005 in order to control the systematic promotion of space development, to manage the launch of space objects and to produce guidelines for handling potential compensation claims for damage caused by space accidents.[76] As space development involves great expenses and a high risk as a national strategic and public industry, the Korean Government had recognized the need to establish the legal basis for this undertaking as well as the establishment of the promotion plan every five years.

In 2004 it proposed a "*Draft for the Korean Space Development Promotion Act*" which was then passed by the State Council of the Korean Government on December 21, 2004 and it was submitted to the National Assembly where it was passed by majority resolution on May 3, 2005.

The Space Development Promotion Act (Hereinafter referred to as Korean Space Act) was then transferred to the Government on May 17, 2005 where it was proclaimed with Law no. 7538 on May 31, 2005.

It came into force six months later, on December 1, 2005. But this Act was amended 12 times for 15 years for by the revision of the Government Organization Act etc. until on March 30, 2020 and this Act composed of 29 Article.[77]

The Korean Space Act was in accordance with Korea's international obligations under the various UN space treaties and conventions such as the Space Treaty of 67,[78] the Rescue Agreement of 1968,[79] the Liability Convention of 1972[80] and the Registration Convention of 1975[81].

The Korean drafters of this Act has studied and analyzed the contents of the foreign space law such as the United States, the United Kingdom, Canada, Russia, Germany, France, Japan, Italy, Sweden, Austria, Australia, Spain, Argentina, Belgium, Brazil, Netherlands and South Africa etc. and then reflected partly the merits and main points of the aforementioned foreign space acts and four space treaties. The Republic of Korea has enacted its Space Development Promotion Act and Act on Compensation for Damage Caused by Space Objects so as to become an excellent model in the countries of Asia-Pacific region.

The main features of the Korean Space Act which comprises 29 articles are as follows:

1. Purpose (Art. 1).
2. Government Responsibilities (Art. 3).
3. Formulation of a Basic Plan for Promoting Space Development (Art. 5).
4. Establishment of National Space Committee (Art. 6).
5. Designation of Institution Specializing in Space Development (Art. 7).
6. Domestic Registration of Artificial Space Objects (Art. 8).
7. International Registration of Artificial Space Objects (Art. 9).
8. Artificial Space Objects and Management of Registers of Artificial Space
9. Objects (Art. 10).
10. Permission to Launch Space Launch Vehicles (Art. 11).
11. Revocation of Permission for Launch, and Hearings (Art. 13).
12. Liability for Damages Caused by Space Accidents (Art. 14).
13. Formulation of Basic Plan for Preparing against Dangers in Space (Art. 15).
14. Designation of Space Environment Surveillance Agency, etc. (Art. 15, 3).
15. Composition, etc. of Space Accident Investigation Committee (Art. 16).
16. Dissemination and Utilization of Satellite Information (Art. 17).
17. Support for Space Development Projects by Private Sector (Art. 18).
18. Requests for Assistance or Cooperation in Space Development (Art. 20)
19. Compensation for Damages Caused by Launched Space Objects (Art. 20-2)
20. Implementation of Space Development Projects Related to National Security (Art. 21)
21. (22) Rescue of Astronauts (Art. 22).
22. (23) Return of Artificial Space Objects (Art. 23).
23. (24) Collection of Data and Surveys of Actual Conditions on Space Development, etc. (Art. 24)
24. (25) Duty of Confidentiality (Art. 25)
25. (26) Penalty Provisions (Art. 27).
26. (27) Administrative Fines (Art.29)

7.7.4. Act on Compensation for Damage Caused by Space Objects

According to the formal request from the Korea Institute of Nuclear Safety on October, 2005, I was made out a *"Tentative Draft for the Act on Compensation for Damage Caused by Space Objects"* and then sent to the aforementioned Korea Institute on December, 2005.

Though *"the Draft for Act on the Compensation for Damage Caused by Space Objects"* (Korean Space Liability Act) was based on Article 14 of the Space Development Promotion Act, this Draft aims at the protection of the victims in the case of space damage caused by space objects and at the quick solution of disputes.

The *"Draft for Act on the Compensation for Damage Caused by Space Object"* was drawn up in the context of Korea´s due considerations with respect to the Liability Convention of 1972 as well as in connection with the study of legislative examples of domestic acts relating to the compensation for space damage in developed countries.

As twelve congressmen proposed the "*Draft for Act on the Compensation for Damage Caused by Space Object*" to the National Assembly on February 5, 2007, it was discussed by the Science, Technology, Information and Telecommunication Committee under the National Assembly on April 12, 2007.

This Act on the Compensation for Damage Caused by Space Objects (우주손해 배상법: 宇宙損害賠償法) was passed by majority resolution of the Korean National Assembly and proclaimed with Law no. 8714 by the Government on December 21 2007.[82]

But this Act was amended five times for ten years by the revision of the Government Organization Act etc. until on March 30, 2020.[83]Finally this Act was enacted by the method of congressman's legislature. It is composed of nine articles. The main key points of this Act can be described as the following:

1. Purpose (Art. 1).
2. Definition (Art. 2)
3. Relations, etc. with International Treaty (Art. 3).
4. Strict Liability and Concentration of Liability, etc. (Art. 4).
5. Maximum Amount of Compensation for Damage (Art. 5).
6. Purchase of Damage Compensation Liability Insurance Policy (Art. 6).
7. Regulatory Review (Art. 9)

7.7.5. Comment on the Korean Three Space Acts

I would like to comment the Korean three Space Acts as the following important six items.

7.7.5.1. Proposal to Define the Term "Space-Worthiness"

Space-worthiness is a term used to describe whether a satellite and spacecraft has been certified as suitable for safe launching and flight such as airworthiness in aviation law and seaworthiness in maritime law. The application of space-worthiness defines the safe condition of satellite or spacecraft and supplies the basis for judgment of the suitability for launching and space flight of that satellite and spacecraft. It is necessary and desirable for us to adopt a new definition of Space-worthiness in order to provide the safety standard for space activities aimed as well as for the prevention of space accidents.

Therefore a legal entity of launching a space objects shall be responsible before and at the beginning of the launch to exercise and check with due diligence with respect to;

1. Make the satellite and spacecraft space-worthy,
2. Ensuring the space worthiness of the satellite, the space launch vehicle or the spacecraft,
3. Providing the proper crew, equipment and supply of the spacecraft.

Furthermore a legal entity launched satellite or spacecraft shall be liable for damage sustained by third parties upon provided that damage was caused by the un-fulfillment of space-worthiness.[84]Any person who desires to obtain a certification of space-worthiness stating that a satellite or spacecraft is to launch and fly shall file application therefore with Minister of Science, ICT and Future Planning of the Korean government.

7.7.5.2. Exclusion of the Limited Liability for Damage Caused by the Wilful- Misconduct

Although it was fruitful for us to accept strict liability for damage according to Art. 4 and 5 of the Korean Space Liability Act, but it is necessary to exclude such limited liability if the damage was caused by wilful-misconduct (gross negligence) of a legal entity having launched the space object by their servants or agents and caused intentionally or recklessly, knowing that damage would probably occur[85].

7.7.5.3. Proposal to Regulate the Joint and Several Liability

If two or more legal entities undertakes a launching, it will be necessary to insert a provision relating to their joint and several liability for compensation in case of a space accident caused by space objects.

7.7.5.4. Change From the Korean Won to Unit of Account in Sum of Compensation for Damage

Most of the international aviation and maritime treaties as well as the Korean Commercial Code, German Commercial Code (Handelsgestzbuch: HGB), the Japanese International Commodity Transport Act and the Civil Aviation Law of the People's Republic of China has been adopted the Unit of Account (SDR) such as sum to be paid in compensation for damage in Unit of Account (Special Drawing Rights: SDR) decided by the currency of UN International Monetary Fund.

Therefore it is greatly necessary to adopt this recognized currency account unit also in the case of a space accident in Korea and Art. 5 of the Korean Space Liability Act should be changed as Unit of Account.

The sums mentioned in terms of Unit of Account shall be deemed to refer to the SDR as defined by the International Monetary Fund under the United Nations. Conversion from Unit of Account to the national currencies shall, in case of judicial proceedings, be made according to the value of such currencies in terms of the SDR at the date of the judgment. [86]

7.7.5.5. Need to Recognize the Right to Receive Preferential Payment

It is necessary to adopt the victim's right to receive preferential payment for funeral expense before the settlement of other claims in case of a space accident.

7.7.6. Proposal on the Establishment of Korea National Space Agency

Korea has been carrying out its space program step by step in accordance with its national space plans. Several accomplishments from 2010 onward marked milestones in Korean space technology development.

Therefore it is greatly desirable for us to establish a new Korea National Space Agency (KNSA: tentative title) in order further efficiently Korean space industry in the future. KNSA could be a similar body to the Japanese Aerospace Exploration Agency (JAXA), the British National Space Centre, the French Centre National d'Etudes Spatiales, the German Aerospace Center,

The Swedish National Space Agency, the Chinese National Space Administration (CNSA) and the Indian Space Research Organization.[87])

It could also be an alternative to establish a Korea National Space Agency which could be a governmental organization or national legal entity and could take a similar function like the Canadian Space Agencies in Canada, Japan,[88]) China, India, Italy,

Israel, Russia and the NASA of USA in order to efficiently develop Korean space industry. From now on Korea anticipates the accomplishment of its space programs actively and will continuously strengthen exchanges and cooperation with all countries in the World under the principle of equality, friendly relations and mutual benefits.

Together with all other peoples around the globe Korea will make due contributions towards the peaceful utilization of space resources and the promotion of human progress and prosperity.[89])

Since its foundation in 1989, KARI has expanded its international cooperation in the aerospace field to 28 organizations in countries including the USA, Russia, the UK, France, China, Israel, etc.

Through collaboration, KARI has improved its R&D capabilities. Korea is also trying to participate in the ISS program. KARI has been in negotiations with NASA over the joint ACCESS mission. At the same time, talks with Boeing/NASA on participation in Zarya module in space shuttle are continuing.

7.7.7. Korean Pathfinder Lunar Orbiter to July, 2022

This will be South Korea's biggest step forward into the world of space exploration. using six different instruments, KPLO (Korean Pathfinder Lunar Orbiter), the country's first lunar probe, will work to map out the presence of natural resources on the moon from orbit.

The KPLO is a planned lunar orbiter by the Korea Aerospace Research Institute (KARI) of the Republic of Korea. The orbiter, its science payload and ground control infrastructure, are technology demonstrators.

The orbiter will also be tasked with surveying lunar resources such as water ice, uranium, Helium-3, silicon and aluminum, and produce a topographic map to help select future lunar landing sites.

The mission is planned to be launched in July 2022 on a Falcon 9 rocket. The Republic of Korea's space agency, KARI together with NASA produced a lunar orbiter feasibility study in July 2014.

Korean Lunar lander and a 20kg rover will be arrived at the moon on 2025.

Figure 18.
Source: Source: http://www.businesskorea.co.kr/news/articleView.html?idxno=38484

The two agencies signed an agreement in December 2016 where NASA will collaborate with one science instrument payload telecommunications, navigation, and mission design. The Korean Lunar Exploration Program (KLEP) is divided in two phases.[90] Phase 1 is the launch and operation of KPLO, which will be the first lunar probe by the Republic of Korea, meant to develop and enhance South Korea's technological capabilities, as well as map natural resources from orbit.

The main objectives of this mission are to enhance the South Korean technological capabilities in the ground and in outer space, and to "increase both the national brand value and national pride".

The specific technological objectives are:

- Development of critical technologies for lunar exploration.
- Produce a topographic map for support to select future lunar landing sites, and to survey lunar resources such as water ice, uranium, Helium-3, silicon, and aluminum.
- Development and validation of new space technologies.

From the lunar science perspective, understanding the water cycle on the Moon is critical to mapping and exploitation. Solar wind protons can chemically reduce the abundant iron oxides present the lunar soil, producing native metal iron (FeO) and a hydroxy1 ion (OH-) that can readily capture a proton to form water (H_2O). Hydroxyl and water molecules are thought to be transported throughout the lunar surface by mysterious unknown mechanisms, and they seem to accumulate at permanently shadowed areas that offer protection from heat and solar radiation.[91]

7.7.8. The Korea Aerospace Research Institute

The Korea Aerospace Research Institute (KARI) established in 1989, is the aeronautics and space agency of South Korea. Its main laboratories are located in Daejeon, in the Daedeok Science Town. KARI's vision is to continue building upon indigenous launch capabilities, strengthen national safety and public service, industrialize satellite information and applications technology, explore the moon, and develop environmentally-friendly and highly-efficient cutting-edge aircraft and core aerospace technology. Current projects include the KSLV-2 launcher.

Past projects include the 1999 Arirang-1 satellite. The Naro Space Center is a South Korean spaceport in South Jeonlla Nam Province, Goheung County, operated by the state-run Korea Aerospace Research Institute.[92] The Naro Space Center is located about 485 km (300 mile) from capital city of Seoul, Korea.

It includes two launch pads, a control tower, rocket assembly and test facilities, facilities for satellite control testing and assembly, a media center, an electric power station, a space experience hall and a landing field. It supported 4 launches, and will also support the KSLV-II launch in 2021, and SSLV launches in 2025.[93]

Phase 1 incorporates the launch and operation of a lunar orbiter called Korea Pathfinder Lunar Orbiter (KPLO), which will be the first lunar probe by South Korea, meant to develop and enhance South Korea's technological capabilities, as well as map natural resources from orbit. Phase 2 will include a lunar obiter, a lunar lander, and a rover to be launched together on a KSLV-II South Korean rocket from the Naro Space Center by 2030.[94]

Figure 19.
Source: https://www.kari.re.kr/eng.do

Figure 20.
Source: https://shindonga.donga.com/3/all/13/111457/1

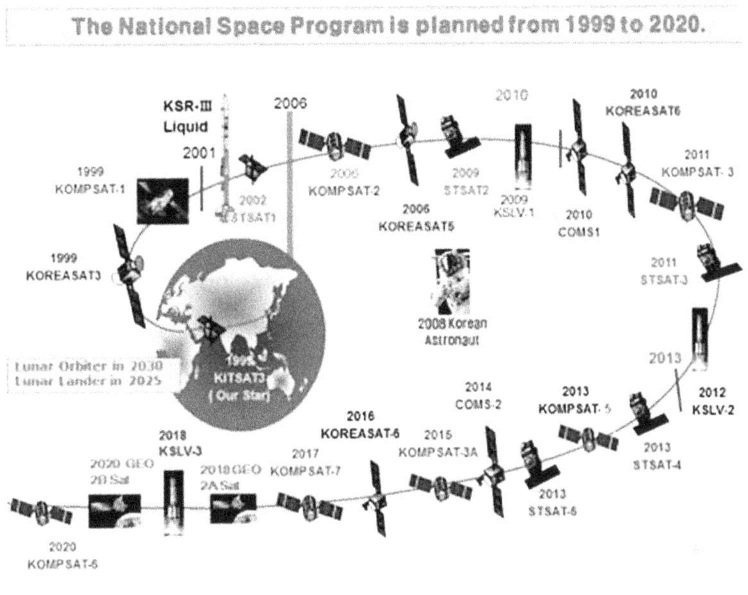

It includes two launch pads, a control tower, rocket assembly and test facilities, facilities for satellite control testing and assembly, a media center, an electric power station, a space experience hall and a landing field. It supported 4 launches, and will also support the KSLV-II launch in 2021, and SSLV launches in 2025.[95]

Phase 1 incorporates the launch and operation of a lunar orbiter called Korea Pathfinder Lunar Orbiter (KPLO), which will be the first lunar probe by South Korea, meant to develop and enhance South Korea's technological capabilities, as well as map natural resources from orbit.

Phase 2 will include a lunar obiter, a lunar lander, and a rover to be launched together on a KSLV-II South Korean rocket from the Naro Space Center by 2030.[96]

ENDNOTES

1 https://spacelaws.com/articles/space-licensing-in-china

2 https://spacelaws.com/articles/space-licensing-in-china

3 https://oxfordre.com/planetaryscience/oso/viewentry/10.1093$002facrefore$002f9780190647926.001.0001$002facrefore-9780190647926-e-66

4 https://oxfordre.com/planetaryscience/oso/viewentry/10.1093$002facrefore$002f9780190647926.001.0001$002facrefore-9780190647926-e-66

5 https://www.spacedaily.com/reports/China_expects_to_introduce_space_law_around_2020_999.html

6 https://www.chinadaily.com.cn/china/2014-11/17/content_18930721.htm

7 http://www.cnsa.gov.cn/english/n6465668/n6465670/c6478978/content.html

8 https://spacenews.com/chinese-commercial-launch-sector-regulations-released-new-launch-vehicle-plans-unveiled

9 https://en.wikipedia.org/wiki/Chang%27e_4

10 https://en.wikipedia.org/wiki/Chang%27e_5

11 *Zhao, Jiannan; Xiao, Long; Qiao, Le; Glotch, Timothy D.; Huang, Qian (27 June 2017). "The Mons Rümker volcanic complex of the Moon: a candidate landing site for the Chang'e-5 mission". Journal of Geophysical Research: Planets.* ***122*** *(7): 1419–1442.*

12 China's Planning for Deep Space Exploration and Lunar Exploration before 2030. (PDF XU Lin, ZOU Yongliao, JIA Yingzhuo. *Space Sci.*, 2018, 38(5): 591-592. doi:10.11728/cjss2018.05.591

13 "*Lunar plans for phase IV*". Archived from the original on 2019-04-15. Retrieved 2019-01-13.

14 Chang'e-4 landing to be a step along a road of lunar exploration for China. Andrew Jones, *Space News*. December 28, 2018.

15 https://www.researchgate.net/publication/335595976_Space_Law_the_Present_and_the_Future

16 http://app.myzaker.com/news/article.php?pk=5e2965518e9f095a6d6f96e7

17 https://www.space.com/22743-china-national-space-administration.html

18 https://en.wikipedia.org/wiki/China_National_Space_Administration

19 China spent 7 times more than India on space; US 13 times". The Times of India. 31 January 2020. Retrieved 1 February 2020. While India's ISRO spent about $1.5 billion in 2018, US' Nasa spent $19.5 billion and China's CNSA spent $11 billion

20 "Archived copy". Archived from the original on 28 February 2008. Retrieved 9 March 2008.CS1 main: archived copy as title

21 https://www.prsindia.org/sites/default/files/bill_files/Draft%20Space%20Activities%20Bill%202017.pdf

22 Water on the Moon – ISRO". www.isro.gov.in. Archived from the original on 1 August 2019. Retrieved 10 September 2019.

23 Thomas, Arun. "Mangalyan". CNN. Archived from the original on 26 September 2014. Retrieved 24 September 2014.

24 ISRO sends record 104 satellites in one go, becomes the first to do so". The Economic Times. Archived from the original on 15 February 2017. Retrieved 15 February 2017.

25 https://en.wikipedia.org/wiki/Chandrayaan-2

26 https://www.isro.gov.in/chandrayaan2-latest-updates

27 https://www.technologyreview.com/s/615291/lunar-missions-before-nasa-2024-artemis-moon-landing-space-x-mining/

28 https://www.isro.gov.in/pslv-c25-mars-orbiter-mission

29 *"Vision and Mission Statements"*. www.isro.gov.in. Archived from the original on 28 March 2019. Retrieved 28 March 2019.

30 Roger D. Launius (2018). The Smithsonian History of Space Exploration: From the Ancient World to the Extraterrestrial Future. Smithsonian Institution. pp. 196–. ISBN 978-1-58834-637-7.

31 Isro's golden jubilee: 50 years of space explorations". Archived from the original on 17 August 2019. Retrieved 21 September 2019.

32 "About ISRO - ISRO". Archived from the original on 28 March 2019. Retrieved 28 March 2019.

33 https://en.wikipedia.org/wiki/National_Institute_of_Aeronautics_and_Space

34 Lapan A3 Satellite to be Launched in June". May 25, 2016.

35 https://www8.cao.go.jp/space/english/index-e.html

36 https://www.spacetechasia.com/an-overview-of-japans-space-activities

37 Outer Space Basic Plan, Cabinet Decision, Apr. 1, 2016 at 24, Cabinet Office website in Japanese.

38 https://thelawreviews.co.uk/edition/the-space-law-review-edition-1/1211969/japan

39 https://www.loc.gov/law/foreign-news/article/japan-two-outer-space-related-laws-enacted

40 https://ispace-inc.com/hakuto-r/eng/about

41 https://www.technologyreview.com/s/615291/lunar-missions-before-nasa-2024-artemis- moon-landing-space-x-mining/

42 The Japan's Mainichi Newspaper on September 28, 2020. Retrieved from https://mainichi.jp/english/articles/20200928/p2g/00m/0fe/058000c

43 https://english.kyodonews.net/news/2020/09/f66797eb75fa-japan-eyes-using-moon-water-as-fuel-for-space-exploration-in-2030s.html

44 "*JAXA plans probe to bring back samples from moons of Mars"*. 10 June 2015 – via Japan Times Online.

45 https://spaceflightnow.com/2020/02/20/phobos-sample-return-mission-enters-development-for-2024- launch

46 NASA confirms contribution to Japanese-led Mars mission. Stephen Clark, *Space Flight Now*. 20 November 2017.

47 https://akatsuki.isas.jaxa.jp/en/mission/

48 https://eos.org/articles/asteroid-mission-attempts-touchdown-sample-grab

49 https://www.japantimes.co.jp/news/2019/01/01/national/global-collaboration-needed-future-space-missions/#.XpTl-Yt7kfg

50 https://global.jaxa.jp/press/2018/06/20180608_mmo.html

51 https://www.isas.jaxa.jp/en/missions/spacecraft/current/mmo.html

52 https://en.wikipedia.org/wiki/Jupiter_Magnetospheric_Orbiter

53 https://en.wikipedia.org/wiki/Jupiter_Icy_Moons_Explorer

54 https://www.japantimes.co.jp/news/2019/01/01/national/global-collaboration-needed-future-space-missions/#.XpTl-Yt7kfg

55 https://www.isas.jaxa.jp/e/forefront/2013/adam/03.shtml

56 http://www.esa.int/Science_Exploration/Space_Science/Cassini-Huygens/Cassini_concludes_pioneering_mission_at_Saturn

57 McCurry, Justin (15 September 2007). *"Japan launches biggest moon mission since Apollo landings"*. guardian.co.uk/science. London. Retrieved 16 September 2007.

58 JAXA - New JAXA Philosophy and Corporate Slogan". Retrieved 12 June 2015.

59 Japan delays launch of unmanned lunar lander to second half of fiscal 2019". The Japan Times. 4 June 2015. Retrieved 22 June 2015.

60 https://www.officialgazette.gov.ph/downloads/2019/08aug/20190808-RA-11363-RRD.pdf

61 "House okays PH Space Dev't bill on 2nd reading". Philippine News Agency. 27 November 2018. Retrieved 29 November 2018.

62 Ronnicque, Neyzielle (15 August 2019). "Philippine Space Agency signed into law". DOST- Advanced Science and Technology Institute. Retrieved 6 September 2019.

63 https://en.wikipedia.org/wiki/Philippine_Space_Agency

64 Philippines ready and able to create its own space agency, minister says". *The Japan Times*. Kyodo. 1 February 2019. Retrieved 3 February 2019.

65 https://en.wikipedia.org/wiki/Korean_Committee_of_Space_Technology

66 https://en.wikipedia.org/wiki/National_Aerospace_Development_Administration#cite_note-2

67 *"North Korea fires projectile into waters off eastern coast"*. Fox News. Archived from the original on June 8, 2013. Retrieved May 19, 2013.

68 https://en.wikipedia.org/wiki/List_of_North_Korean_missile_tests

69 *"Missiles of North Korea"*. Missile Threat. Retrieved March 28, 2020.

70 https://csis-prod.s3.amazonaws.com/s3fs-public/publication/161108_AHC_korean_missile_Forces.pdf

71 http://www.kcna.co.jp/item/2012/201203/2012-03-28ee.html

72 https://en.wikipedia.org/wiki/National_Aerospace_Development_Administration

73 https://elaw.klri.re.kr/kor_service/lawView.do?hseq=46909&lang=ENG

74 Doo Hwan Kim, *Essay for the Study of the International Air and Space Law*, December 2008,rean Studies Information Co. Ltd, pp. 372-375.

75 Kim Doo Hwan, *The Problems on the Space Exploitation Program and Legislation in Korea,* The Korean Journal of Air and Space Law (Vol. 18, 2003, pp.218~223); According to the invitation from the KARI, I had lectured with my topic entitled the *Necessity of Legislating for the Development and Promotion of the Korean Space Industry* on June 23, 2003.

76 Doo Hwan Kim, *"Necessity for Enacting the Space Law in Korea.* Japanese Academic Journal Kiyo; 4(1), Proceedings of the Research Institute of Social Systems of Chuogakuin University, Chiba-ken, Japan, at 39-52.

77 https://elaw.klri.re.kr/kor_service/lawView.do?hseq=46397&lang=ENG

78 Treaty on Principles Governing the Activities of States in the Exploration and Use of Outer Space, Including the moon and other Celestial Bodies.

79 Agreement on the Rescue of Astronauts, the Return Astronauts and the Return of Objects Launched into Outer Space.

80 Convention on International Liability for Damage Caused by Space Objects.

81 Convention on Registration of Objects Launched into Outer Space.

82 金斗煥、「韓国に於ける新しい宇宙宇宙開發振興法と宇宙損害賠償法試案の主な內容及び 將來の課題」、(紀要第6巻第2号)、2006年3月10日)、日本中央学院大学社会システム研究所発行、124-126頁.

83 https://elaw.klri.re.kr/kor_service/lawView.do?hseq=46485&lang=ENG

84 Legislative Examples could be provided by the International Convention for the Unification of Certain Rules of Law relating to Bills of Lading (1924) Art. 3, the Korean Commercial Code, Art. 794 (Seaworthiness and the Japanese Commercial Code), Art. 738 (Seaworthiness, The Korean Aviation Act), Article 15 (Certificate of Airworthiness, The Japanese Aviation Act), Article 10 (Certificate of Airworthiness).

85 In this connection international and national legislative examples could be: Art. 25 of the Warsaw Convention of 1929, Art. 13 of the Hague Protocol of 1955, Art. 22 of the Convention for the Unification of Certain Rules for International Carriage by Air (Montreal Convention of 1999), Art. 4 of the Convention on Limitation of Liability for Maritime Claims of 1976, Art. 8 of the United Nations Convention on the Carriage of Goods by Sea (Hamburg Rule of 1978), Art. 21 of the United Nations Convention on International Multimodal Transport of Goods of 1980, Art. 769, 797 of the Korean Commercial Code of 2007, Art. 13 Para. 2 of the Japanese International Commodity Transport Act of 1992.

86 In this connection international and national legislative examples could be: Art. 2 of the Montreal Additional Protocol No.1, No.2, No.3 and Art. 22 of the Montreal Protocol of 1975, Art. 22-23 of the Montreal Convention 1999, Art. 6 of the Convention on Limitation of Liability for Maritime Claims of 1976, Art. 6 of the Hamburg Rule of 1978, Art. 18 of the United Nations Convention on International Multimodal Transport of Goods of 1980, Art. 770, 797 of the Korean Commercial Code of 2007, Art. 13 of the Japanese International Commodity Transport Act of 1992, Art. 45-46 of the German Air Transport Act (Luftverkehrgesetz of 2007), Art. 129 of the Civil Aviation Law of the People's Republic of China.

87 Doo Hwan Kim, *"Example Legislation on the Space Relations of Every Countries in the World and Main Contents of the Space Exploration Promotion Act and Future Task in Korea"*, The Korean Journal of Air and Space Law, Vol.20, No.1, (June 2005), at 35-36.

88 Article 25~30 of the Japanese Basic Space Law of 2008.

89 http://www.cnsa.gov.cn/main_e.asp

90 Korean Lunar Exploration Program. Korean Aerospace Research Institute (KARI). Accessed on 25 January 2019.

91 South Korea's 2018 Lunar Mission. Paul D. Spudis, *Air and Space Magazine*. 26 September 2016.

92 https://en.wikipedia.org/wiki/Korea_Aerospace_Research_Institute

93 https://en.wikipedia.org/wiki/Naro_Space_Center

94 https://en.wikipedia.org/wiki/Korea_Aerospace_Research_Institute

95 https://en.wikipedia.org/wiki/Naro_Space_Center

96 https://en.wikipedia.org/wiki/Korea_Aerospace_Research_Institute

Chapter 8
The European Countries' National Legislation on the Outer Space Laws and Organizations

ABSTRACT

This chapter looks at legislation that European countries have placed on outer space.

8.1. AUSTRIA

8.1.1. Austrian National Space Legislation

Austria has ratified all five United Nations space treaties. In order to implement the obligations contained in these treaties, Austria has enacted the Austrian Outer Space Act in 2011, which was complemented by the Outer Space Regulation in 2015.

Furthermore, the Austrian Outer Space Act established the Austrian Registry for Space Objects.[1]

On 28 December 2011 the Austrian Federal Law on the Authorization of Space Activities and the Establishment of a National Registry (Austrian Outer Space Act) entered into force after its publication in the Federal Law Gazette I No 132.

The planned launch of two Austrian satellites (Brite Austria Mission Tugsat 1 & UniBRITE) raised the question whether the existing legal framework in Austria was sufficient to deal with the specificities of space activities.

The Austrian entities involved felt that there was a need to accompany the project with all its legal aspects. For this purpose, they approached the Austrian National Point of Contact for Space Law (NPOC) of the European Centre for Space Law (ECSL) which had been founded in 2001 by Prof. Christian Brünner at the University of Graz, Austria. [2]

DOI: 10.4018/978-1-7998-7407-2.ch008

8.1.2. Contents of Austrian Outer Space Law

With its only 17 articles, the Outer Space Law is relatively short but is meant to be a comprehensive act which deals with all the legal aspects connected to space activities, as the following;

1. authorization,
2. supervision,
3. mitigation of space debrid,
4. registration,
5. liability,
6. insurance,
7. transfer of the space object,
8. recource,
9. enforcement and sanctions.

The relatively short length corresponds to the hitherto modest independent Austrian space activities, primarily dedicated to science, research and education. Nevertheless, the act is also designed for commercial space activities which might become more important in the future and, this is rather unusual, also for governmental space activities.

In the drafting and negotiation process, under the lead of the Federal Ministry there was a general desire among the different ministries involved to establish a transparent system and a guaranteed flow of information between the different federal and territorial entities as far as prospective space activities are concerned.

Furthermore, building up of the relevant knowhow as regards, in particular, authorization and supervision was regarded as a challenge that could best be met by covering all kinds of Austrian space activities in the future.

Particular emphasis has been put on the mitigation of space debris. Small satellites are a very attractive option for newcomers as they are becoming technically more accessible and cheaper. Nevertheless, the responsibility of the States in this respect is more pertinent than ever before.

States have the duty to avoid that small satellite projects, as laudable and welcome they are, do not harm large and expensive space activities which are of vital interest to the world population at large.Thus, the new Austrian Outer Space Act covers a variety of issues connected to both international obligations of Austria in accordance with the five UN space treaties but also takes into account the particular needs of space activities being carried out by Austrian operators which for the moment encompass mainly educational and research purposes.

8.1.3. NPOC Austria and the European Centre for Space Law

The National Point of Contact for Space Law Austria (NPOC) Austria acts as an interface between the European Centre for Space Law (ECSL) and persons interested in space law in Austria (ECSL).

It is coordinated by Prof Irmgard Marboe at the Department of European, International and Comparative Law of the University of Vienna and supported by the Austrian Research Promotion Agency and the Austrian Ministry for Transport, Innovation and Technology (About NPOC).

Its main objective is the promotion and development of space law in Austria.

This includes a variety of activities such as the organization of conferences, the elaboration of space law-related publications, the teaching of space law as well as the publication of an annual Space Law Newsletter (Activities).

In addition, the NPOC informs about recent space law-related developments (News) and upcoming events (Events) and provides information on legal texts and literature relating to the exploration and use of outer space as well as a link collection on space in general (Services).[3]

8.2. BELGIUM

8.2.1. Belgian Space Law

The Law of 17 September 2005 (Belgian Official Journal of 16 November 2005) on the Activities of Launching, Flight Operation or Guidance of Space Objects entered into force on 1st January 2006. It has been revised by the Law of 1 December 2013 (entered into force on the day of its publication in the Belgian Official Journal, on 15 January 2014). The Belgian space law aims particularly at:

- ensuring the legal and material safety of operational space activities Performed under Belgian jurisdiction;
- developing an appropriate legal framework for the hosting of this sector in Belgium.

This legislation puts Belgium at the forefront of the space faring countries.

It notably allows operators to settle down in Belgium and to perform their activities under Belgian jurisdiction (satellite operation, space navigation,...). Such activities are subject to the authorization of the Minister for Science Policy who may impose peculiar conditions aiming at ensuring everyone's safety as well as the interests of all participants.

The revision allowed to better circumscribe the scope of the law by providing a more precise definition of the notions of space object, of operator including in the specific case of non-maneuverable space objects.

The Law of 17 September 2005 on the Activities of Launching, Flight Operation or Guidance of Space Objects composed of 21 articles and seven chapters as the following.

Chapter I, General provisions,
Chapter II, Authorization and Supervision of the Activities,
Chapter III, Content of the Dossier and Procedure,
Chapter IV, Transfer of activities,
Chapter V, The national Register of Space Objects,
Chapter VI, Liabilities, counterclaims and measures in the event of falling space objects
Chapter VII, Final provision.

The Royal Decree implementing some provisions of the Law was adopted on 19 March 2008 and entered into force at the time of its publication in the Official Journal, on 11 April 2008. This legislation puts Belgium at the forefront of the space faring countries.

The revision allowed to better circumscribe the scope of the law by providing a more precise definition of the notions of space object, of operator including in the specific case of non-maneuverable space objects.

8.2.2. Belgian National Register of Space Objects

This law also establishes a National Register of Space Objects which should make Possible the registration by Belgium of satellites or other spacecraft's of which it would be the Launching State, according to the provisions of the UN space treaties. Finally, the law organize between the Belgian State and the operator, a system of sharing of the liability for the damage caused by the space object.

8.3. DENMARK

8.3.1. The Danish Outer Space Act

The Danish Outer Space Act of 2016 entered into force on 1 July 2016 (Act no. 409 of 11 May 2016). The Danish Executive Order on requirements in connection with approval of activities in outer space, etc. entered into force on 1 July 2016 (Executive Order no. 552 of 31 May 2016).

The Danish Outer Space Act composed of 23 Articles and 10 Part as the followings;

Part 1, Scope and objective
Part 2, Definitions
Part 3, Approval of space activities
Part 4, Disclosure obligations, changes and withdrawal
Part 5, Registration
Part 6, liability and insurance
Part 7, Transfer of space objects or space activities
Part 8, Supervision, exemption and authorization
Part 9, Penalties
Part 10, Entry into force and transitional provisions, etc.

The Danish Outer Space Act and Executive Order on requirements in connection with approval of activities in outer space, etc. is the legal basis for Danish space activities. The Danish Outer Space Act and Executive Order apply to space activities carried out within the Danish State.

8.3.2. Outline of the Danish Outer Space Act

The Danish Outer Space Act and Executive Order on requirements in connection with approval of activities in outer space, etc. is the legal basis for Danish space activities. Pursuant to section 6 (1) no. 4 of the Act, approval of space activities requires that the operator has taken appropriate measures with regard to space debris management.

As a general rule, within 25 years of the end date of the functional operating period of the space object, the space object must either safely leave its earth orbit again or safely be placed into an orbit where it is deemed not to constitute a danger to other space activities.

The Danish Outer Space Act and Executive Order apply to space activities carried out within the Danish State. Furthermore, the Act and Executive Order apply to space activities carried out outside the Danish State on Danish craft or facilities or by Danish operators.

8.3.3. Relation to the International Mechanisms

The European Cooperation for Space Standardization (ECSS) and the International Organization for Standardization (ISO) are explicitly referred to in the Danish Executive Order on requirements in connection with approval of activities in outer space, etc. Other relevant standards and guidelines such as The Space Debris Mitigation Guidelines of the Committee on the Peaceful Uses of Outer Space, the IADC Space Debris Mitigation Guidelines, the ITU Recommendation ITU-R S.1003 and European Code of Conduct for Space Debris Mitigation can also be taken into consideration.[4]

8.3.4. Space Travel in Denmark

Compared to its size, Denmark has a strong presence among the space faring nations. From design and delivery of spacecraft components through universities and private companies, to the first Danish astronaut Andreas Mogensen, who flew on the International Space Station in 2015. Danish involvement in space travel is boosted by its role as a founding member of the European Space Agency (ESA).

It is our hope that Denmark's commitment to space and space exploration may be strengthened further, allowing for further opportunities for the Danish space industry, as well as increasing the public awareness of space, and desire for adventures into the unknown.

8.4. FINLAND

8.4.1. Act on Space Activities of Finland

The Act on Space Activities of Finland entered into force on 23 January 2018. The Act is supplemented by the Decree of the Ministry of Economic Affairs and Employment on Space Activities. The objective of the Act is to provide space activities carried on in Finland with a clear framework, smooth authorization process and uniform operator obligations.

A predictable and legally clear operating environment promotes the competitiveness and growth of and a safe and secure operating environment for space industry and helps attract new actors and investments to Finland. The starting point is a positive approach towards space activities.

Finland is committed to the UN treaties on outer space. Under the treaties, Finland is responsible for national space activities and is obliged to supervise them. The Act implements in national law the international obligations binding on Finland under the UN treaties on outer space. [5]

This Act composed of 22 Section and four Chapters as following;

Chapter 1, General provisions,
Section 4, Definition
Chapter 2, Operator's obligations,
Section 5, Authorization for space activities and conditions for authorization Space
Section 7, Liability for damage and the State's right of recourse
Section 8, Obligation to insure
Section 9, Safe conduct of space activities
Section 10, Environmental protection and space debris
Section 11, Transfer of space objects and space activities to others
Chapter 3, Changes and supervision
Section 12, Obligation to provide information
Section 13, Amending and withdrawing an authorization
Section 14, Supervision
Section 15, Right of inspection
Chapter 4, Miscellaneous provisions
Section 16, Found space objects
Section 17, Opinions
Section 18, Right of an authority to obtain information
Section 20, Appeal
Section 21, Penal provisions
Section 21, Entry into force
This Act enters into force on 23 January 2018.

8.4.2. Decree of the Ministry of Economic Affairs and Employment on Space Activities

Decree of the Ministry of Economic Affairs and Employment on Space Activities, to be adopted and entering into force 23rd January 2018. According to the act, space activities means launching space objects into outer space, operation and control of space objects in outer space and returning space objects to the Earth.

This decree composed of nine Sections as the following;

Section 1, Applying for authorization
Section 2, Application for authorization
Section 3, Space debris
Section 4, Registry of space objects
Section 5, Insurance
Section 6, Transfer of space objects or space activities
Section 7, Obligation to provide information
Section 8, Annual report
Section 9, Entry into force

8.5. FRANCE

8.5.1. French Space Operation Law

The 2008 French space operation law provides that every operator has to carry out, for any space operation, an impact assessment on the environment, and a hazard study with a plan to manage risks and ensure safety of populations, properties, public health and the environment.

The adoption of the 2008 French space operation law or "*LOI du 3 juin 2008 relative aux opérations spatiales)* " on June 3, 2008, which addressed several issues intimately linked to the privatization of space activities, marked the outcome of several years of discussions, and finally provided France with a legal framework for activities in outer space. The French space operation law composed of thirty articles and eight title as the following;

Title 1, Definitions
Title 2, Authorization of Space Operations
Chapter 1, Operations to Space Authorizations
Chapter 2, Conditions for Granting Authorizations
Chapter 3, Obligations of Authorizations Holders
Chapter 4, Administrative and Penal Sanctions
Title 3, Registration of Launched Space Objects
Title4, Liability
Chapter 1, Liability Towards Third Parties
Chapter 2, Liability Towards Persons Taking Part in the Space Operation
Title 5, Provisions Relating to the Research Code
Title 6, Intellectual Property
Title 7, Space-Based Data
Title 8, Transitory and Final Provisions

This highly expected piece of legislation, adopted a couple of days before the beginning of the French Presidency of the Council of the European Union, aimed at setting out the legal framework for French space activities, and at the same time, it clearly indicated the perspectives of France in terms of business, national sovereignty and independence.

The 2008 French space law sets up an authorization and continuous supervision process of the outer space activities of the French operators, in accordance with the international treaties, and in particular, the 1967 Outer Space Treaty and the 1972 Liability Convention.

This process allows mastering the liability of France for space activities for which it is responsible, in accordance with the aforementioned international treaties.

The 2008 French space law provides that every operator has to carry out, for any space operation, an impact assessment on the environment, and a hazard study with a plan to manage risks and ensure safety of populations, properties, public health and the environment.

The authorization process and the assessment of compliance with the Technical Regulation provides assurance that the operators have the means, resources, necessary skills and are appropriately organized to perform the operation in compliance with the 2008 French space law.

It also allows competent authorities to verify that compliance is maintained throughout operational life of the space object up until disposal, through the processing of the technical and organizational events. For the sake of consistency, it was decided to extend previously established *ad hoc* practices and allocate them a legal origin so to ensure their dissemination in the space industry. This firstly concerns the cross waiver of liability clauses.

This practice consists in considering that each party in a space related contract shall individually bear the risks of the activities performed and renounce to all claims against the other party (*Article 20 of the 2008 French space law*).

The second noteworthy piece of codification carried out in the 2008 French space law concerns the liability ceiling for launch operators, and the mandatory insurance obligations for amounts below the ceiling (*Article 6 of the 2008 French space law*).

The 2008 French space law also provides an insurance obligation for risks in orbit.[6]

8.5.2. Space Debris Mitigation Guidelines

The 2008 French space operation law leads to authorize and supervise all the space operations performed by French operators, and takes into account the long-term development of space activities.

In particular, the Technical Regulation was developed with due consideration paid to the *Space Debris Mitigation Guidelines* adopted by the Committee on the Peaceful Uses of Outer Space and endorsed by the United Nations General Assembly, the recommended practices and voluntary guidelines proposed by the *Inter-Agency Space Debris Coordination Committee* (IADC), and the Committee on Space Research (COSPAR), as well as the existing international technical standards, including those published by the International Organization for Standardization (ISO), and the Consultative Committee for Space Data Systems (CCSDS), generally accepted by the international space community for the safe conduct of outer space activities.

8.5.3. The French Space Agency

The National Centre for Space Studies (CNES) (French: *Centre national d'études spatiales*) is the French government space agency (administratively, a "public administration with industrial and commercial purpose").

The French Space Agency, CNES, founded in 1961, is a major actor for outer space sustain ability and implementation of voluntary guidelines. CNES is the government agency responsible for shaping and implementing France's Space policy in Europe.

Its headquarters are located in central Paris and it is under the supervision of the French Ministries of Defence and Research.

It operates from the Toulouse Space Center and Guiana Space Centre, but also has payloads launched from space centers operated by other countries. CNES is member of Institute of Space, its Applications and Technologies.

As of April 2018[update], CNES has the second largest national budget- €2.438 billion- of all the world's civilian space programs, after only NASA.[7]

Through its ability to innovate and its forward-looking vision, CNES is helping to foster new technologies that will benefit society as a whole, focusing on five themes: access to outer space (with *Ariane*), telecommunications, observation, science, security and defense. Its numerous affiliations and participations in companies, CNES was in a position which enabled the control to take place.

8.5.4. Guiana Space Centre

Guiana Space Centre is Europe's main the launch facility, owned by ESA (European Space Agency) and operated by CNES (the French space agency). Located near Kourou in French Guiana, on the northern coast of South America at 5.2° N, 52.8° W, it is ideal for launching satellites into geosynchronous orbits and also supports missions into polar orbit.

It benefits from Earth's rotation near the equator, which gives a free boost of 460 meters per second to eastward-launched. Originally selected by CNES, the Kourou site has been vastly expanded with funding from ESA for its Ariane launcher programs.

Figure 1. Guiana Space Centre located near Kourou in French Guiana on the northern coast of South America
Source: https://www.daviddarling.info/encyclopedia/G/Guiana_Space_Centre.html

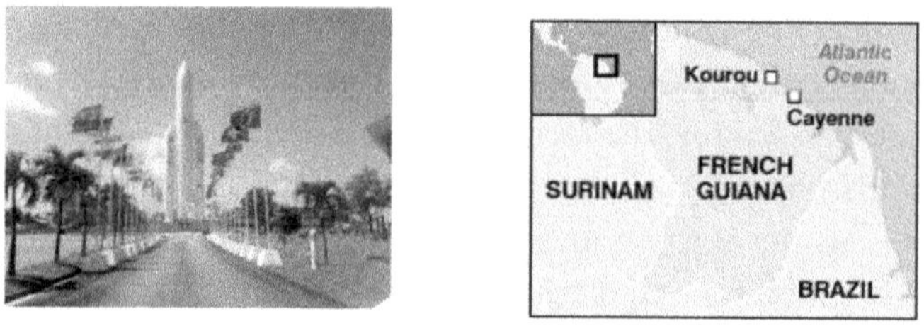

The Ariane 5 facilities alone cover 21 square kilometers out of a total of 850 square kilometers occupied by the spaceport.[8]

Europe's Spaceport was first opened 14 April 1964, meaning that this spring 2019 marks the 55th birthday of the building of the Spaceport.

The Ariane 5 facilities alone cover 21 square kilometers out of a total of 850 square kilometers occupied by the spaceport.[9]

Europe's Spaceport was first opened 14 April 1964, meaning that this spring 2019 marks the 55th birthday of the building of the Spaceport.

The area that houses the ESA office was constructed in the 1960s, and then added to as necessary; I was actually present when the last building projects were executed in 1996. Today, there are a myriad of buildings and, as you can imagine, maintenance and general upkeep is no negligible task.[10]

Figure 2. Guiana Space Centre at the Northern Coast of South America
*Source: https://www.daviddarling.info/encyclopedia/G/Guiana_*

8.6. GERMANY

8.6.1. Law Governing the Transfer of Administrative Functions in the Sector of Outer Space Activities

The Law governing the transfer of administrative functions in the sector of outer space activities (Gesetz zur Übertragung von Verwaltungsaufgaben auf dem Gebiet der Raumfahrt: Raumfahrtaufgabenübertragungsgesetz) was promulgated by German Federal Government on 22 August 1998.

The abovementioned Law composed of 4 Section as the following;

Section 1, Transfer of administrative functions, The supreme federal authorities in charge of outer space activities confer to the German Aerospace Center (DLR) the authority to perform administrative functions in the sector of outer space activities in its own name and acting as an administrative authority under public law.
Section 2, If the DLR passes on budgetary funds as part of the administrative functions performed by it, these funds should be transferred to it for management.
Section 3, Right of Audit of the Federal Audit Office
Section 4, Entry into force

8.6.2. The Satellite Data Security Act

The Parliament (Bundestag) passed the "Act to Safeguard the Security Interests of the Federal Republic of Germany from Endangerment by the Distribution of High-Grade Earth Remote Sensing Data" *(Satellitendatensicherheits-gesetz–Satellite Data Security Act- SatDSiG).* The law became effective on December 1, 2007.

The purpose of this Act is, on the one hand, to safeguard the security and foreign policy interests of the Federal Republic of Germany in connection with the distribution and commercial marketing of satellite-acquired earth remote sensing data especially on international markets.

On the other hand, the Act will create legal certainty for affected companies and make the terms of operating in the new business areas calculable for the developing companies involved in satellite data marketing - thus also for the expanding geo-data industry.

It will therefore fulfil an important condition, enabling German companies to translate satellite applications into commercially viable business models and enter new sales markets. The Act on Satellite Data Security is needed to provide legal certainty, establish binding rules, and ensure enforcement; it is also necessary in the interest of sending a foreign policy signal and furnishing legal foundations for the issues addressed.

The *SatDSiG* became necessary since highly capable space-based earth remote sensing satellites are constructed in Germany with the intention of the worldwide commercial marketing of the acquired images/data.[11]

The Federal Parliament (Bundestag) has passed the "*Satellite Data Security Act*" as the following:

Part 1, Section 1, Scope of Application
Section 2, Definition of Terms
Part 2, Section 3, Operation of high-grade earth remote sensing system
Section 4, Operator license requirements
Section 7, Obligation to provide information
Section 8, Rights of entry and inspection
Section 9, Measures of the responsible authorities
Part 3, Dissemination of data
Part 4, Priority compliance with requests from the Federal Republic of Germany
Part 5, Implementing regulations
Section 24, Responsibility
Section 26, Fees and expenses
Part 6, Fine Provisins, Penal Provisions
Section 28, Administrative offenses
Section 29, Criminal offenses
Section 30, Offenses committed in foreign countries by German citizens
Part 7, Transitional and final Provisions.

Earth remote sensing data generated by high grade Earth Observations Systems (EOS) is made progressively available for worldwide civilian commercialization.

These data have a grade of quality, which was previously produced only by classified military and intelligence service satellites and that was used exclusively in closely defined environments. The distribution of these high value or high grade earth remote sensing data may endanger foreign or security policy interests. It is therefore of vital interest to establish efficient means to secure the control of these satellites and the distribution of their data products.

The German Satellite Data Security Act (Satellitendatensicherheitsgesetz, SatDSiG) became necessary since highly capable space-based earth remote sensing satellites are constructed in Germany with the intention of the worldwide commercial marketing of the acquired images/data. The act provides legal certainty, establishes binding rules, and ensures their enforcement.[12]

8.6.3. PTS: ALINA, 2021

German company Planetary Transportation Systems (PTS) is not in good shape. The company recently filed for bankruptcy protection amidst bad losses in recent years. It hasn't given up on ALINA, its lunar lander prototype. This won't be a lengthy mission—PTS just wants to prove its technologies work and it can get to the moon in one piece. But ALINA will include two rovers built by Audi, and a miniature biosphere to grow plants.[13]

8.6.4. The German Aerospace Center

The German Aerospace Center (German: *Deutsches Zentrum für Luft- und Raumfahrt e.V.*), abbreviated DLR, is the national center for aerospace, energy and transportation research of Germany. Its headquarters are located in Cologne and it has multiple other locations throughout Germany. The DLR is engaged in a wide range of research and development projects in national and international partnerships.

In addition to conducting its own research projects, DLR also acts as the German Space Agency.

As such, it is responsible for planning and implementing the German space programme on behalf of the German Federal Government. As a project management agency, DLR also coordinates and answers the technical and organizational implementation of projects funded by a number of German federal ministries. DLR has approximately 8,200 employees at 20 locations in Germany.[14]

Its institutes and facilities are spread over 13 sites, as well as offices in Brussels, Paris and Washington D.C. DLR has a budget of €1 billion to cover its own research, development and operations.

Approximately 49% of this sum comes from competitively allocated third party funds (German: *Drittmittel*). In addition to this, DLR administers around €860 million in German funds for the European Space Agency. In its capacity as project management agency, it manages €1.279 billion in research on behalf of German federal ministries.[15]

8.7. ITALY

8.7.1. Italian Space Law

"The Law for the Implementation for the Convention on International Liability for Damages Caused by Space Objects (Law No 23)" was enacted on January 25, 1983 in Italy.

This Law governs compensation for damage caused by space objects launched by foreign states that are parties to the 1972 Liability Convention. The other key Italian legislation is Law No 153 of July 12, 2006: "Registration of Objects launched into Outer Space."

Under this law, Italian Space Agency maintains the national registry and collects information required by the 1975 Registration Convention.[16] The "Law on the Establishment of the Italian Space Agency" is Law No. 186 of 30 May 1988, The Italia Space Agency was established by the aforementioned law.

8.7.2. Italian Space Agency

The Italian Space Agency (Italian: Agenzia Spaziale Italiana; ASI) is a government agency established in 1988 to fund, regulate and coordinate space exploration activities in Italy. The agency cooperates with numerous national and international entities who are active in aerospace research and technology. [17]

Internationally, the ASI provides Italy's delegation to the Council of the European Space Agency and to its subordinate bodies as well as representing the country's interests in foreign collaborations.

ASI's main headquarters are located in Rome, Italy, and the agency also has direct control over three operational centers.

The Centre for Space Geodesy (CGS) located in Matera in Italy. As well as these ASI has access to its own spaceport, the Broglio Space Centre (formerly the *San Marco Equatorial Range*) on the coastal sublittoral of Kenya, currently used only as a communications ground station.

In 2016 ASI's annual revenues budget was approximately €1.6 billion and directly employed around 200 workers.[18]

8.8. LUXEMBOURG

8.8.1. Space Resources Law

The *Law on the Exploration and Use of Space Resources* (the Space Resources Law) adopted by the Luxembourg Parliament on 13 July 2017, as promogulated by Government on July 20, 2017 and then entered into force from 1 August 2017, [19]creates a licensing and supervisory regime in Luxembourg addressing the ownership of natural resources acquired in space.

This Space Resources Law composed of 18 Articles. Similar to the *2015 US Commercial Space Launch and Competitiveness Act, the Space Law* provides that commercial companies operating within its regulatory framework may legally appropriate natural resources acquired in space from celestial bodies known as Near Earth Objects (NEOs).

Luxembourg is the first European country to adopt legislation regulating the ownership of natural resources acquired in space by commercial companies, providing legal certainty for commercial projects in the space sector.

The 1967 Outer Space Treaty (OST) ratified by 109 countries including Luxembourg and the 1979 Moon Agreement (MA) ratified and accessed by 18 countries, established principles for the peaceful and free exploration of space by nation states.

However, OST and MA does not address the ownership by private organizations of the resources harvested from NEOs by, for example, asteroid mining, including metals, minerals, and gases. [20]

The Law on the Exploration and Use of Space Resources states that "space resources are capable of being appropriated," thus giving private companies ownership of space resources that they extract. (Art. 1) Under this Law, Luxembourger corporations or European companies that have their registered office in Luxembourg may extract space resources for commercial use after obtaining approval from the Government of Luxembourg. (Art. 2-4)

Anyone who explores and uses space resources without government authorization is subject to between eight days and five years of imprisonment and/or a fine of €5,000 to €1,250,000 (about US$5,900-US$1,480,000). (Art. 18)

The Government of Luxembourg will evaluate a company's application in light of a number of criteria. Upon receiving the application, the ministers in charge of the economy and space activities will review it to make sure that the company has:

(A) its central administrative body and its registered office in Luxembourg,
(B) the financial, technical, and legal means planned out and implemented in order to explore and use space resources, and
(C) a reliable internal governance scheme. (Art. 7)

In this process, the company must disclose the identities of its largest shareholders or associates who have the power to influence the company, so that the ministers in charge may evaluate their reputation, knowledge, skills, financial soundness, and risk or history of money laundering or terrorist financing. (Art. 8) The reputation, knowledge, skills, experience, and conduct of the members of the company's management body are also taken into account. (Art. 9)

Moreover, the ministers will verify that the annual accounts of the company are audited by one or more experienced and approved company auditors (Art. 11) and that the company has submitted a risk assessment of the space mission with the proof of financial means to cover any risks. (Art. 10)

Once a company has been granted approval to conduct exploration and use of space resources, the authorization is and non-assignable. (Art. 5) Therefore, another party cannot carry out an exploration mission through or for the authorized company. (Art. 2)[1]

8.8.2. The Luxembourg Space Agency

Lux innovation has been supporting the Luxembourg Space Agency (LSA) since its inception. Luxembourg is a catalyst for collaboration, technological innovation and the commercial development of space, bringing together the expertise and financing required to create a sustainable space economy of the future. Today, the Luxembourg space sector's contribution to the nation's GDP is amongst the highest in Europe.

In line with its determination to create a solid, vibrant and innovative space sector, Luxembourg has developed an overarching space policy.

The objectives are to develop the Luxembourg space ecosystem and create synergies with businesses and organizations outside the space sector, encourage the development of key skills and expertise, and develop Luxembourg and its space sector internationally.

For more than three decades, Luxembourg has been at the forefront of commercial and co-operative initiatives that have shaped a vibrant space economy.

In 2005, Luxembourg joined the European Space Agency (ESA) as a full member, laying the foundation for participation by the Grand Duchy's companies in ESA R&D programmes and space exploration projects. Today Luxembourg is one of the top five per capita contributors to ESA. [21]The most recent step taken by Luxembourg's Ministry of the Economy was the launch in 2016 of the Space Natural Resources, initiative, positioning the country as a pioneer in the space-based resources economy.

In the coming years, the exploration and utilization of space resources is set to generate attractive opportunities in Luxembourg and worldwide for established and start-up players in fields including materials science, additive manufacturing, remote sensing, communications, robotics, data analytics and artificial intelligence.

Figure 3. Moon Village made in Luxembourg Space Agency
Sources: https://space-agency.public.lu/en.html

Launched in February 2016 and led today by the Luxembourg Space Agency (LSA), the Space Resources. Initiative positions Luxembourg as a pioneer in the exploration and utilization of space resources.

The identification and utilization of space resources is fast becoming a reality, driven by a revolution in space technology, accelerating exploration of outer space and the eventual scarcity of certain resources on Earth.

Building on its long history at the forefront of the commercial satellite communications industry, Luxembourg aims to play a leading role in the exploration and utilization of these resources. Its goal is to ensure that space resources explored under its jurisdiction serve a peaceful purpose, are gathered and used in a sustainable manner compatible with international law and for the benefit of humankind.

Luxembourg's vision is built on support for advanced research activities and technological capabilities, drawing on the country's existing expertise in the space sector and its ongoing strategy of economic diversification into future-oriented high-tech industries.

The Luxembourg Space Agency plays a significant role in promoting Luxembourg internationally and strengthening international collaboration. The LSA represents Luxembourg in the European Space Agency (ESA) and space-related programmes of the European Union.

It contributes to and supports United Nations activities relating to space matters, especially with regard to the Space Resources. Partnerships between Luxembourg and other countries on space activities and space resources are also expanding.[22]

8.9. NORWAY

8.9.1. Outer Space Act

"Act on launching objects from Norwegian territory etc. into outer space(No. 38)" entered into force on June 13, 1969. I would like to introduce briefly the contents of the Norwegian Outer Space Act as the following;

1. Without permission from the Norwegian Ministry concerned, it is forbidden to launch any object into outer space from:
    a. Norwegian territory, also including Svalbard, Jan Mayen and the Norwegian external territories.
    b. Norwegian vessels, aircrafts etc.
    c. Areas that are not subject to the sovereignty of any state, when the launching is undertaken by a Norwegian citizen or person with habitual residence in Norway. Certain terms can be set for such permission as described in paragraph one.
2. The Ministry can issue regulations on control etc. of activities as described in 1.3.
3. This act enters into force immediately.

8.9.2. The Norwegian Space Agency

The Norwegian Space Agency (NOSA,[23] formerly the Norwegian Space Centre, NSC; Norwegian*: Norsk Romsenter*) is a that follows up Norway's public space activities. Established in 1987 when Norway joined the European Space Agency (ESA), NOSA is an agency of the Norwegian Ministry of Trade and Industry.

The goal is for space activities to be of use for society and contribute to business development. NOSA is also charged with safeguarding and promoting Norway's interests in relation to ESA and the EU space programmes and in bilateral agreements with other countries.

In brief, NOSA is to ensure that Norway benefits as much as possible from any space activity in which the Norwegian state becomes involved.

The agency is located in Oslo and the nearest commuter train station is Skøyen station. Space Norway is a related Norwegian space company, established in 2014.

While it operates as a commercial limited liability company, it is entirely governmentally owned by the Government of Norway. Space Norway has contracted with SpaceX for a 2022 launch of its two-satellite Arctic Satellite Broadband Mission (ASBM) system.[24]

8.10. SPAIN

8.10.1. The Royal Decree on Spanish Registry of Space Objects

"Royal Decree (No, 278/1995 of 24 February 1995 Prime Minister's Chancellery) on the Space Exploration, Establishment in Spain of the Registry of Objects Launched into Outer Space" was composed of seven articles.

Following the accession of Spain on 20 December 1978 to the Convention on Registration of Objects Launched into Outer Space, adopted by the United Nations General Assembly on 12 November 1974 (Official State Gazette No. 25 of 29 January 1979) (RCL 1979, 269 and ApNDL 8191), appropriate internal measures must be adopted to enable Spain to comply with the terms of that Convention, especially with regard to the establishment of a registry of objects launched into outer space and to the notifications that have to be made to the Secretary-General of the United Nations.[25]

In Spain, a Royal Decree of 1995 was established national registry for space objects which should be maintained by the Ministry of Foreign Affairs.

It provides that entry shall be made in the Spanish registry in respects of space objects that have been launched or whose launching has been procured by the Spanish State, or that have been from Spain or Spanish Facility. [26]

I would like to introduce the contents of the Royal Decree on the Space Registry such as the following main articles;

Article 1, The Spanish Registry of Objects Launched into Outer Space, hereinafter referred to as the "Spanish Registry", shall be established.

Article 2, The Spanish Registry shall be kept by the Division of Multilateral Economic Relations and Development of the Department of International Economic Relations of the Ministry of Foreign Affairs.

Article 4, The term "space object" includes component parts of a space object as well as the launch vehicle and parts thereof.

Article 6, The registration of each space object shall contains five items.

8.10.2. The National Institute for Aerospace Technology

The National Institute for Aerospace Technology (Instituto Nacional de Técnica Aeroespacial: INTA) is an autonomous research agency of the Government of Spain responsible for the state aerospace, aeronautics, hydrodynamics, and defense and security technologies research. The INTA was established in 1942, as the *Instituto Nacional de Técnica Aeronáutica* (National Institute of Aeronautics), and it was integrated in the Ministry of the Air.

It has its headquarters in Torrejón de Ardoz, near Madrid. Its budget, €190 million in 2019, comes from the Spanish Ministry of Defence and from its own projects with the industry. As of 2017 INTA had a total of 1500 employees, 80% of them are dedicated to R&D activities.

Its two main areas of activity are research and development (for example, in propulsion, materials, remote sensing) and certification and testing (for example, software, metrology).

8.11. SWEDEN

8.11.1. Act on Space Activities

The Swedish Act on Space Activities (1982: No. 963) composed of Section six as the following main Articles.[27]

Section 1. This Act applies to activities in outer space (space activities).
Section 2. Space activities may not be carried on from Swedish territory by any party other than the Swedish State without a license.
Section 3, A license to carry on space activities is granted by the Government.
Section 4. A license may be withdrawn if the conditions of the license have been ㅜ disregarded or if there are other particular reasons for it.
Section 5. Any person who willfully or through negligence carries on space activities without the necessary license shall be sentenced to a fine or to imprisonment for at most one year.

The Decree on Space Activities (1982: No.1069) in Sweden composed of four Sections. Sweden established its Act on Space Activities together with a Decree on Space Activities in 1982.

The Act encompasses all space 'activities carried on entirely in outer space and, in addition, the launching of objects into outer space and all measures to maneuver or in any other way affect objects launched into outer space.

Anyone carrying on space activities (as defined by the Swedish Act) and any Swedish natural or legal persons carrying on space activities anywhere else requires license from the Swedish government with this comprehensive scope, Sweden asserts jurisdiction over nationals with respect to space activities.

Both the potential, international responsibility of Sweden and its potential international liability with respect to privately conducted space activities are thus deal with. [28]

The Decree accompanying the Space Activities Act established a National Board for Space Activities as the government authority for granting license and maintaining the national register of launch space objects.

8.11.2. The Swedish National Space Agency

The Instituto Nacional de Técnica Aeroespacial (INTA, National Institute for Aerospace Technology) is an autonomous research agency of the Government of Spain responsible for the state aerospace, aeronautics, hydrodynamics, and defense and security technologies research. The INTA was established in 1942, as the Instituto Nacional de Técnica Aeronáutica (National Institute of Aeronautics), and it was integrated in the Ministry of the Air. It has its headquarters in Torrejón de Ardoz, near Madrid.

Its budget, €190 million in 2019, comes from the Spanish Ministry of Defense and from its own projects with the industry. As of 2017, INTA had a total of 1500 employees, 80% of them are dedicated to R&D activities.[29]

The Swedish space programme is mostly carried out through international cooperation. Out of a yearly budget of approximately 900 Mkr (100 M€), about 70% is used to support ESA programmes of importance to Sweden. The programme has included a sequence of satellite missions, both national ones and in cooperation with other nations.[30]

8.12. THE NETHERLANDS

8.12.1. The Dutch Space Activities Act

The Dutch "Rules Concerning Space Activities and the Establishment of a Registry of Space Objects (Space Activities Act)"[31] composed of twenty eight Sections and seven Chapters as the following;

CHAPTER 1. GENERAL PROVISIONS
CHAPTER 2. LICENCES
§ 1. License for space activities
Section 3, 1. It is prohibited to perform space activities as referred to in Section 2 without a license issued for this purpose by Our Minister
§ 2. License application
Section 4, License application shall be submitted to Our Minister.
Section 5, 1, Our Minister will decide on a license application within six months after having received it.
§ 3. Disasters
Section 8, 1, 1. The license is not transferable.
CHAPTER 3. REGISTRY OF SPACE OBJECTS
Section 11, 1, Our Minister shall maintain a registry with Information concerning space objects that are being used in connection with space activities as referred to in Section 2.
CHAPTER 4. REDRESS
Section 12, 1, If the State is obliged to pay compensation under Article VII of the Outer Space Treaty or the Liability Convention, the State is entitled to recover this sum, in full or in part, from the party whose space activity has caused the damage.
CHAPTER 5. ENFORCEMENT
Section 16,1, Our Minister shall not impose an administrative penalty if the infringement cannot be imputed to the perpetrator.
Section 17, 1, The power to impose an administrative penalty will lapse five years after the infringement is committed.
Section 22,1, An administrative penalty shall be paid within six weeks after the decision imposing the penalty has come into force.
CHAPTER 6. AMENDMENTS TO OTHER LEGISLATION
CHAPTER 7. CONCLUDING PROVISIONS
Section 27, This Act shall enter into force on a date to be determined by Royal Decree.
Section 28, This Act may be cited as: the Space Activities Act.

The Dutch Decree of 13 November 2007, containing rules with regard to a registry of information concerning space objects (Space Objects Registry Decree) composed of eleven Articles for procedure of registry on space objects.

8.12.2. The Netherlands Space Office

The Netherlands Space Office (NSO) is the space agency of the Netherlands. Three Dutch government ministries - the Ministry of Economic Affairs and Climate Policy; the Ministry of Education, Culture and Science; and the Ministry of Infrastructure and Water Management - along with the Netherlands Organization for Scientific Research (NWO), an independent administrative body under the auspices of the Dutch-government Ministry of Education, Culture and Science, signed an agreement in October 2008 for the establishment of the NSO.

It was established as of 1 July 2009 following the merger of the space activities of the Netherlands Agency for Aerospace Programmes (NIVR), which was then disestablished on January 2011 and several other institutes (KNMI, SRON).[32]

The NSO was established by the Dutch government to advice on the Netherlands Space Policy and to develop and administer the country's space programme. NSO represents The Netherlands to international space organizations, including the multi-country-member European Space Agency (ESA); the United States National Aeronautics and Space Administration (NASA) and the Japan Aerospace Exploration Agency (JAXA). NSO is also the central government point of contact for the space community in the Netherlands.

The NSO also promotes education and communication on space, especially the Netherlands space activities.

8.13. THE UNITED KINGDOM

8.13.1. The 1986 Outer Space Act

The Outer Space Act had enacted in the United Kingdom on July 18, 1986.

This Act composed of 4 Sections 15 and 15 items as the following.[33]

Section 1, Application of Act
- (1) Activities to this Act applies.
- (2) Persons to whom this Act applies.

Section 2. (3) Licensing of activities
- (4) Prohibition of unlicensed activities.
- (5) Grant of licence.
- (6) Charges
- (7) Terms of licence.
- (8) Transfer, Variation, suspension or termination of license

Section 3. Other controls
- (9) Register of space objects.
- (10) Power to give directions
- (11) Obligation to indemnify government against claim

Section 4. General
- (12) Regulation
- (13) Offences.
- (14) Minor definitions.
- (15) Index of defined expressions.[34]

The United Kingdom (UK) is one of the pioneering nations in space activity. Today it has a leading reputation in a number of key space sectors, both in the scientific and commercial arenas. The UK is also a signatory to four UN space treaties, and takes seriously its international commitment to the safe, responsible, and sustainable use of space; only with this global commitment can the world continue to exploit and enjoy the unique opportunities that space offers.

The Outer Space Act 1986 ("OSA") is the legal means by which the UK regulates the use of outer space.

The Outer Space Act is the basis for the regulation of activities in outer space carried out by organizations or individuals established in the United Kingdom, or in one of its overseas territories or Crown dependencies.

It confers licensing and other powers on the Secretary of State for Business, Energy and Industrial Strategy, who acts through the UK Space Agency to exercise these powers.[35]

The Outer Space Act seeks to:

- ensure compliance with the UK's various obligations under international treaties band principles covering the use of outer space, including liability for damage caused by space objects,
- the registration of objects launched into outer space and the principles for the remote sensing of the Earth,
- ensure that space activities licensed by the UK do not undermine national security
- manage the risk of claims for third-party damage being brought against the UK Government, and to transfer some of that liability from the UK Government and taxpayers to the licensed organization or individual whose space activities caused the third-party damage.

8.13.2. The 2015 Amendment to the Outer Space Act

8.13.2.1. Outline

Section 12 of the Deregulation Act 2015 amended the Outer Space Act to make it obligatory for a license to specify the maximum amount of a licensee's liability to indemnify the Government in respect of activities authorized by the license.

The 1986 Act was amended in 2015, to introduce a limit to the operator's indemnity to the UK Government for third-party claims brought against the UK.

Act has also been extended, through Orders in Council, to apply with modifications to Crown dependencies and overseas territories. Fees regulations have also been made under powers in the Act.

8.13.2.2. Applying for a License

UK nationals and UK companies intending to launch or procure the launch of a space object, operate a space object, or carry on any other activity in outer space should make themselves familiar with the provisions of the Outer Space Act 1986, plus the amendments made to the Act by the Deregulation Act 2015.

Unless acting as an employee or agent of another organization, you need to apply for a license at least six months in advance of carrying out the licensable activity. In certain circumstances it may be possible to process an application in a reduced timescale, although no guarantees can be given. Applications should be made using the license application form below.

There is flexibility in the UK approach to licensing and the UK Space Agency encourages potential applicants to contact them as early as possible to discuss the best way forward and solution for their mission.[36]

8.13.2.3. The UK's Space Liability and Indemnity Regime

Under section 10 of the Outer Space Act 1986 (OSA), operators must indemnify the UK Government for claims brought against the latter other than in circumstances set out in that section. The operator's indemnity to the UK Government is a vital part of the UK's overall space risk management approach, helping the UK offset some of the liability that it incurs on behalf of satellite operators.

Up until 2015, this indemnity was unlimited: if a claim was brought against UK Government, the Government could seek recovery from the operator, however large the claim. The Outer Space Act was amended by the Deregulation Act 2015.

Henceforth, all licenses issued must state a limit to the operator's liability to indemnify UK Government for claims made against the latter.

This means that, if a third-party claim is brought against the UK Government, the operator would be liable to indemnify the Government up to a limit, and the UK Government would meet the remaining liability.

The limit on the liability to indemnify the Government, ("the Indemnity limit") also referred to as the indemnity 'cap', is set out in license conditions and determined by the UK Space Agency on a case-by-case basis.

The UK Space Agency has full discretion to vary the indemnity limit for each license, depending on the risks associated with that mission. It should be noted that each satellite is licensed individually and so in effect the indemnity limit is set on a per-satellite basis.

8.13.2.4. Third-Party Liability Insurance

in order to better protect both operators and the UK Government from third-party claims, the UK Space Agency may also require licensees to hold third-party liability (TPL) insurance for regulated activities, which currently consist of in-orbit operations and the procurement of overseas launches. The UK Government should be named on the insurance policy as an additional insured. The minimum insurance cover required is determined at the discretion of the UK Space Agency; operators are free to hold more than the UK Space Agency requires.[37]

For launch, in the majority of cases involving single satellite missions employing established launchers, this insurance cover would be limited to € 60 million. On the 1st October 2018, the UK Space Agency introduced a new approach to its in-orbit third-party liability insurance requirements.

The changes were introduced in response to feedback from the space industry and in recognition of a rapidly diversifying space sector. Such a diversification of technology and risk meant that the previous model of requiring € 60 million TPL insurance in all cases was no longer appropriate.[38]

8.13.2.5. The 2019 Space Industry Act

The Space Industry Act of 2018 is a major step towards establishing a flourishing space industry in the United Kingdom, with a modern, safe and supportive regulatory framework. Work across government is now ongoing, at pace, to develop the detailed regulations to implement the Act. Licenses for spaceflight and associated activities from the UK under the Act cannot be issued until these supporting regulations have been approved by Parliament.

These licenses will be issued by either the UK Space Agency (UKSA) or the Civil Aviation Authority (CAA) depending on the activity. Licenses for spaceflight and associated activities from the UK under the Act cannot be issued until these supporting regulations have been approved by Parliament. These licenses will be issued by either the UK Space Agency (UKSA) or the Civil Aviation Authority (CAA) depending on the activity. Further information is set out below but we recommend that all potential applicants read 'Understanding The Space Industry Act 2018' and take their own legal advice:

The 2018 Space Industry Act requires the licensing of any launch, procurement of launch, operation or return of vehicles capable of operating above the stratosphere (approximately 50km) and any activities that take place above the stratosphere.

It also requires the licensing of any balloon capable of operating within the stratosphere (approximately 10km-50km) with people aboard. These are collectively called operator licenses.

In this document, an operator license involving launch is generically referred to as a launch vehicle operator license but would be either a launch vehicle operator license for sub-orbital activities or a launch vehicle operator license for space activities.

In most cases, the Civil Aviation Authority (CAA) will regulate sub-orbital activities and the UK Space Agency (UKSA) will regulate space activities.

The exception is that the UKSA will regulate all spaceflight activities involving the use of rockets, regardless of whether they are involved in sub-orbital or space activities.

Rockets not involved in spaceflight activities (i.e. those that are not capable of operating above the stratosphere) will continue to be regulated by the CAA under Article 96 of the Air Navigation Order 2016 (the ANO). The CAA also remain responsible for airspace under aviation law. The CAA has a world class reputation for safe and efficient regulation of UK airspace through the well-established process of

Airspace Design Change.[39] The CAA supports the creation of a competitive market for UK spaceflight and will make airspace design decisions in accordance with established processes and statutory duties for safety, security and the environment. Safety is at the heart of the Space Industry Act.

An important consideration for regulators is how license applicants will be asked to demonstrate that the risks these activities pose to the uninvolved general public (AKA third parties) are as low as reasonably practicable and at a level that is acceptable to the regulator. The Space Industry Act is composed of 72 Items, 9 Sections and 12 Schedules as the following.[40]

Section 1. Regulation of Space Flight etc. (Item 1~4)
Section 2. Rang Control (Item 5~7)
Section 3. License (Item 8~15)
Section 4. Exercise of regulatory functions by bodies other than Secretary of State (Item 16)
Section 5. Individuals taking part in spaceflight activities etc (Item 17~18)
Section 6. Safety (Item 19~21)
Section 7. Security (Item 22~25)
Section 8. Enforcement etc. (Item 26~33)
Section 9. Liabilities, Indemnities and Insurance (Item 34~38)
Section 10. Powers in relation to land: Supplementary (Item 39~50)
Section 11. Offences and civil sanctions (Item 51~59)

Section 12. Appeals (Item 60)
Section 12. Miscellaneous (Item 61~66)
Section 12. General (Item 67~72).

8.13.3. The UK Space Agency

The United Kingdom Space Agency (commonly known as the UK Space Agency or UKSA) is an executive agency of the Government of the United Kingdom, responsible for the United Kingdom's civil space programme. It was established on 1 April 2010 to replace the British National Space Centre (BNSC) and took over responsibility for government policy and key budgets for space exploration;[41] it represents the United Kingdom in all negotiations on space matters. The Agency "[brings] together all UK civil space activities under one single management". It is based at the former BNSC headquarters in Swindon, Wiltshire.

Prior to the creation of the Agency, the space and satellite industry in the UK was valued at £9 billion and supported 68,000 jobs.[42] The 20-year aim of the Agency is to increase the industry to £40 billion and 100,000 jobs,[43] and to represent 10% of worldwide space products and services (increasing from the current 6%).

On 30 November 2018, it was announced that the UK Space Agency will abandon ties to the EU's Galileo satellite system following Brexit. Instead, the Agency will manage the United Kingdom's own system of navigation satellites In November 2019, a trade association estimated the cost of this programme at £5bn.[44]

ENDNOTES

1 https://www.loc.gov/law/foreign-news/article/luxembourg-law-on-use-of-resources-in-space-
2 https://www.spacelaw.at/austrian-outer-space-act
3 https://www.spacelaw.at
4 https://ufm.dk/en/legislation/prevailing-laws-and-regulations/outer-space/outer-space-act.pdf
5 https://tem.fi/en/spacelaw
6 https://www.spacelegalissues.com/analysis-of-the-2008-french-space-law/
7 https://en.wikipedia.org/wiki/CNES
8 https://www.daviddarling.info/encyclopedia/G/Guiana_Space_Centre.html
9 https://www.daviddarling.info/encyclopedia/G/Guiana_Space_Centre.html
10 https://blogs.esa.int/spaceport/2019/04/19/happy-55th-birthday-to-europes-spaceport
11 https://www.bmwi.de/Redaktion/DE/Downloads/S-T/satdsig-hintergrund-en.pdf?__blob=Publication File&v=1
12 https://www.bsi.bund.de/EN/Publications/TechnicalGuidelines/TR03140/BSITR03140.html
13 https://www.technologyreview.com/s/615291/lunar-missions-before-nasa-2024-artemis- moon-moon-landing-space-x-mining
14 https://en.wikipedia.org/wiki/German_Aerospace_Center
15 https://en.wikipedia.org/wiki/German_Aerospace_Center#Research_aircraft_2
16 Rams. Jakhu, Joseph N.Pelton, *Global Space Governance: An International Study*, 2,017, 97
17 https://www.asi.it/en/#divTechnology

18 https://en.wikipedia.org/wiki/Italian_Space_Agency
19 https://spacenews.com/luxembourg-adopts-space-resources-law/
20 https://www.ogier.com/news/the-luxembourg-space-law
21 https://space-agency.public.lu/en/agency/international-collaboration.html
22 https://space-agency.public.lu/en/agency/international-collaboration.html
23 Romsenter, Norsk. *"Name change at the Norwegian Space Centre"*. Norwegian Space Agency. Retrieved 2019-03-05.
24 https://en.wikipedia.org/wiki/Norwegian_Space_Agency
25 https://www.unoosa.org/oosa/en/ourwork/spacelaw/nationalspacelaw/spain/royal_decree_278_1995E.html
26 Frans von der Dunk with Fabio Tronchetti, *Handbook of Space Law,* 2015, 171 p.
27 https://www.unoosa.org/oosa/en/ourwork/spacelaw/nationalspacelaw/sweden/decree_on_space_activities_1982E.html
28 Bruno dr Padirac, *"The International Dimensions of Cyber Space Law"*, 2017.
29 https://en.wikipedia.org/wiki/Instituto_Nacional_de_T%C3%A9cnica_Aeroespacial
30 https://en.wikipedia.org/wiki/Swedish_National_Space_Agency
31 https://www.unoosa.org/oosa/en/ourwork/spacelaw/nationalspacelaw/netherlands/space_ activities_actE.html
32 https://en.wikipedia.org/wiki/Netherlands_Space_Office
33 https://www.legislation.gov.uk/ukpga/1986/38/section/6/enacted
34 https://www.legislation.gov.uk/ukpga/1986/38/contents
35 https://www.gov.uk/guidance/apply-for-a-license-under-the-outer-space-act-1986
36 https://www.gov.uk/guidance/apply-for-a-license-under-the-outer-space-act-1986
37 https://www.gov.uk/guidance/apply-for-a-license-under-the-outer-space-act-1986
38 https://www.gov.uk/guidance/apply-for-a-license-under-the-outer-space-act-1986
39 https://www.gov.uk/guidance/applying-for-a-future-licence-under-the-space-industry-act#contents
40 https://www.legislation.gov.uk/ukpga/2018/5/contents
41 The United Kingdom Space Agency (Transfer of Property etc. Order 2011". National Archives, via Legislation.gov.uk. Retrieved 15 December 2015.
42 Hui, Sylvia (24 March 2010. *"British Version of NASA to Begin Work in April"*. Southern Illinoisan. Carbondale, Illinois. Associated Press. p. 6B – via Newspapers.com.
43 New space agency and new international space centre for UK". BNSC. 23 March 2010. Archived from the original on 27 March 2010
44 UKspace 2020 Manifesto". UK Space. 21 November 2019. Retrieved 3 December 2019.

Chapter 9
North America's National Legislation on the Outer Space Laws and Organizations

ABSTRACT

The chapter looks at North American legislation on outer space.

9.1. CANADA

9.1.1. Canadian Space Agency Act

Though Canadian Space Agency Act, S.C. 1990, c. 13 as amended in 2005, but last amended on June 17, 2019. I would like to introduce briefly that Canadian Space Agency Act composed of 29 Articles and table of main contents of it as followings;

Article 1 - Short Title
Article 2 - Interpretation
Article 3 - Establishment of Agency
Article 4 - Objects and Functions There is hereby established an agency of the Government of Canada, to be known as the Canadian Space Agency.
Article 6 - Powers, Duties and Functions of the Minister
Article 12 - Organization of Agency
Article 23 - Annual Report
Article 24 to 27 - Consequential Amendments
Article 28 - Transitional
Article 29 - Coming into Force

DOI: 10.4018/978-1-7998-7407-2.ch009

9.1.2. Canadian Remote Sensing Space Systems Act

Canadian Remote Sensing Space Systems Act, S.C. 2005, c. 45 as amended in 2007.

The Canadian Remote Sensing Space Systems Act and its Regulations came into force on 5 April 2007. The Canadian Remote Sensing Space Systems Act is a national law adopted by Parliament. Under the Canadian Remote Sensing Space Systems Act, no person in Canada shall operate a remote sensing space system in any manner, directly or indirectly, except under the authority of a License.

I also introduce briefly that Canadian Remote Sensing Space Systems Act has composed of 47 Articles and table of main contents of it as followings;

Article 1 - Short Title
Article 2 - Interpretation
Article 3 - Designation of Minister
Article 4 - Application of Act
Article 5 – Operation of Remote Sensing Space System
Article 7 - Applications, Licences and Related Matters
Article 10 - Amendment, Suspension and Cancellation of Licences
Article 14 - Interruptions of Service
Article 15 - Priority Access
Article 16 - Transfer of Remote Sensing Satellites
Article 17 - Inspection
Article 19 - Requests for Information
Article 20 - Regulations
Article 21 - Delegation
Article 22 - No Liability
Article 23 - Administrative Monetary Penalties
Article 38 - Offences
Article 45.1 - Review and Report
Article 46 - Coordinating Amendment
Article 47 - Coming into Force

The main findings of the review are as follows:

1. Although the Act and the Regulations are not perfect, they nevertheless address some very important issues that need to be taken into account when conducting satellite remote sensing activities;
2. The impact of the Act and Regulations on technological development is mixed;
3. The Act and the Regulations are extremely helpful in facilitating compliance with Canada's international agreements and treaties; and,
4. There is a general lack of clarity as to the scope of application of the Act and
5. Regulations. In line with the foregoing findings, this article proposes a handful of carefully considered recommendations aimed at improving the operational efficiency of the Act and the Regulations.[1]

9.1.3. Canadian Space Agency

The Canadian Space Agency (CSA; French: *Agence spatiale canadienne*, ASC) is the national space agency of Canada, established in 1990 by the *Canadian Space Agency Act*. The agency reports to the federal Minister of Innovation, Science, and Economic Development.

The CSA's headquarters are located in Longueuil, Quebec. The agency also has offices in Ottawa, Ontario, at the David Florida Laboratory, and small liaison offices in Houston; Washington, D.C. and Paris.

The CSA has been a cooperating state of the European Space Agency (ESA) since the 1970s[2] and has several formal and informal partnerships and collaborative programs with space agencies in other countries, such as NASA, ISRO, JAXA, and SNSA. The Canadian space program is administered by the Canadian Space Agency. Canada has contributed technology, expertise and personnel to the world space effort, especially in collaboration with ESA and NASA. Canada's contribution to the International Space Station is the $1.3 billion Mobile Servicing System.

9.2. THE UNITED STATES OF AMERICA

9.2.1. The American Space Law

U.S. space law is codified in several laws beginning with the 1958 National Aeronautics and Space Act (NASA Act), signed into law on July 29, 1958. The NASA Act created the National Aeronautics and Space Administration (NASA) to conduct a U.S. civilian space program, with military space activities assigned to the Department of Defense (DOD). The Act has been amended several times over the decades. Other important U.S. laws relating to civil and commercial space, codified in Title 51 of the U.S. Code, include:

- The 1984 Commercial Space Launch Act (P.L. 98-575) that designated the Department of Transportation (DOT) as the federal agency responsible for facilitating and regulating commercial space launch activities (a task currently assigned within DOT to the Federal Aviation Administration's Office of Commercial Space Transportation), [3]
- its 1988 amendments (P.L. 100-657) that provided for government indemnification of commercial space launches for amounts between $500 million and $2 billion (which was extended in several subsequent laws),
- its 2004 amendments that provide for regulation of commercial human spaceflight
- the 2015 Commercial Space Launch Competitiveness Act (see below) makes additional updates/modifications.
- The 1992 Land Remote Sensing Policy Act (P.L. 102-555), which repealed an earlier law (the 1984 Land Remote-Sensing Commercialization Act) and established a regime for facilitating and regulating commercialization of land remote sensing satellites while returning responsibility for the Landsat system to the government. Oversight of commercial remote sensing satellites is assigned to the Department of Commerce and its National Oceanic and Atmospheric Administration (NOAA).

- The 1998 Commercial Space Act (P.L. 105-303), which, inter alia, gave the Department of Transportation regulatory authority over commercial spacecraft that return from space ("reenter"), as well as launches into space.
- The 2005 National Aeronautics and Space Administration Authorization Act (P.L. 109-155) Among the provisions of this act, passed while Congress was under Republican control, was an endorsement of President Bush's 2004 Vision for Space Exploration. The law emphasizes, however, that NASA's programs should be balanced among space science (which uses robotic spacecraft to study the universe and explore the solar system), human spaceflight (President Bush's directive to return humans to the Moon by 2020 and some send them to Mars "and beyond"), and aeronautics. This Act also established the International Space Station as a "national laboratory" to emphasize that its use is not limited to NASA-sponsored research.
- Another important provision is language similar to what was included originally in a 1982 authorization for the Department of Defense (the "Nunn-McCurdy" provision) that sets cost overrun and schedule delay thresholds requiring DOD to notify Congress and take other actions. [4]
- Congress adopted a similar (but not identical) approach for NASA in this 2005 NASA authorization act. NASA must report to Congress if any major NASA development program exceeds baseline development costs by 15% or if a schedule milestone is likely to be delayed by 6 months or more. If it exceeds the baseline cost by 30% or more, not only must Congress be notified, but no funds may spent on it after 18 months from the time of that notification unless Congress reauthorizes the program.
- The 2008 National Aeronautics and Space Administration Authorization Act (P.L. 110-422) This act, passed while Congress was under Democratic control, also endorsed the Vision for Space Exploration with the same stipulation that NASA's programs be balanced among science, human space flight and aeronautics.
- The 2010 National Aeronautics and Space Administration Authorization Act (P.L. 111-267). This act endorsed in part and modified in part President Barack Obama's proposal to significantly change how the United States conducts its human spaceflight program.
- The 2015 Commercial Space Launch Competitiveness Act (P.L. 114-90). This act makes a number of changes to U.S. commercial space policy, including granting property rights to U.S. companies that mine resources from asteroids. It also extends the "learning period" during which the FAA may not issue new commercial human spaceflight regulations until 2023 and extends third party indemnification until 2025.
- Title I is the Spurring Private Aerospace Competitiveness and Entrepreneurship (SPACE) Act, and some refer to it that way. Chapter 513—Space Resource Commercial Exploration and Utilization, Sec. 51303. Asteroid resource and space resource rights, A United States citizen engaged in commercial recovery of an asteroid resource or a space resource under this chapter shall be entitled to any asteroid resource or space resource obtained, including to possess, own, transport, use, and sell the asteroid resource or space resource obtained in accordance with applicable law, including the international obligations of the United States.'' A report submitted by the US government in 2016 states that the US Federal Aviation Administration (FAA) should oversee lunar exploration and asteroid resource mining activities.
- The 2017 NASA Transition Authorization Act (P.L. 115-10), which focuses on continuity at NASA as a presidential transition takes place. Another important space-related law is the FY2020

National Defense Authorization Act (P.L. 115-92), which creates a sixth military service, the Space Force, as part of the U.S. Air Force.[5]

9.2.2. The NASA Lunar Gateway

The Lunar Gateway is an in-development space station in lunar orbit intended to serve as a solar-powered communication hub, science laboratory, short-term habitation module, and holding area for rovers and other robots.[6] It is expected to play a major role in NASA's Atemis Programme. While the project is led by NASA, the Gateway is meant to be developed, serviced, and utilized in collaboration with commercial and international partners.

It will serve as the staging point for both robotic and crewed exploration of the lunar south pole, and is the proposed staging point for NASA's Deep Space Transport concept for transport to Mars.[7]

The science disciplines to be studied on the Gateway are expected to include planetary science, astrophysics, Earth observations, heliophysics, fundamental space biology, and human health and performance.[8]

Gateway development includes all of the International Space Station partners: ESA, NASA, Roscomos, JAXA, and CSA. Construction is planned to take place in the 2020s.[9]

The International Space Exploration Coordination Group (ISECG), which is composed of more than 14 space agencies including all major ones, has concluded that Gateway will be critical in expanding a human presence to the Moon, Mars, and deeper into the Solar System.[10] Formerly known as the Deep Space Gateway (DSG), the station was renamed Lunar Orbital Platform-Gateway (LOP-G) in NASA's 2018 proposal for the 2019 United States federal budget. When the budgeting process was complete, US$332 million[11] had been committed by Congress to preliminary studies.[12] For supporting the first crewed mission to the station (Artemis 3) planned for 2024, the Gateway will be a minimalistic station composed of only two modules: the Power and Propulsion Element (PPE) and the Habitation and

Logistics Outpost (HALO).[13] The Lunar Gateway is planned to be deployed in a highly elliptical seven-day near-rectilinear halo orbit (NRHO) around the Moon, which would bring the station within 3,000 kilometers (1,900 mi) of the lunar north pole at closest approach and as far away as 70,000 kilometers (43,000 mi) over the lunar south pole.[14]

Figure 1. NASA'S Deep Space Gateway & SLS Rocket Kerbal Space Program
Sources: https://www.youtube.com/watch?v=jwVSalth7K4

Traveling to and from columnar space (lunar orbit) is intended to develop the knowledge and experience necessary to venture beyond the Moon and into deep space. NASA will lead an innovative and sustainable program of exploration with commercial and international partners to enable human expansion across the solar system and to runback to Earth new knowledge and opportunities. " President Donald Trump has asked NASA to accelerate our plans to return to the Moon and to land humans on the surface again by 2024.[15]

9.2.3. Landing Again by NASA's Lunar Plan to 2024[16]

We will go with innovative new technologies and systems to explore more locations across the surface than was ever thought possible. This time, when we go to the Moon, we will stay. And then we will use what we learn on the Moon to take the next giant leap - sending astronauts to Mars. NASA's science, technology and human exploration activities touch every aspect of our lives here on Earth and we want to extend our presence to the farthest corners of the universe. In doing so, we will maintain America's leadership in space. Former President Donald *Trump has asked NASA to accelerate our plans to return to the Moon.* [17]

An artist's conception shows the human landing system that's being developed by Blue Origin and its industry partners in the foreground, and Blue Origin's Blue Moon cargo lander in the far background.

Figure 2.
Source: https://www.geekwire.com/2020/jeff-bezos-blue-origin-venture-fleshes-plans-cargo-delivery-moon-2023

Working with U.S. companies and international partners, NASA will push the boundaries of human exploration forward to the Moon for this program. As a result of Artemis, NASA will be able to establish a sustainable human presence on the Moon by 2028 to uncover new scientific discoveries, demonstrate new technological advancements, and lay the foundation for private companies to build a lunar economy. With our goal of sending humans to Mars, Artemis is the first step to begin this next era of exploration.

9.2.4. NASA's Exploration Program on Mars

Mars Exploration Program (MEP) is a long-term effort to explore the planet Mars, funded and led by NASA. Formed in 1993, MEP has made use of orbital spacecraft, landers, and Mars rovers to explore the possibilities of life on Mars, as well as the planet's climate and natural resources. The program is managed by NASA's Science Mission Directorate by Doug McCuistion of the Planetary Science Division As a result of 40% cuts to NASA's budget for fiscal year 2013, the Mars Program Planning Group (MPPG) was formed to help reformulate the MEP, bringing together leaders of NASA's technology, science, human operations, and science missions.[18]

In a little more than seven minutes in the early afternoon of Feb. 18, 2021, NASA's Mars 2020 rover will execute about 27,000 actions and calculations as it speeds through the hazardous transition from the edge of space to Mars' Jezero Crater.

Figure 3. NASA's exploration campaign: back to the moon and Mars
Source: https://moon.nasa.gov/news/46/nasas-exploration-campaign-back-to-the-moon-and-on-to-mars

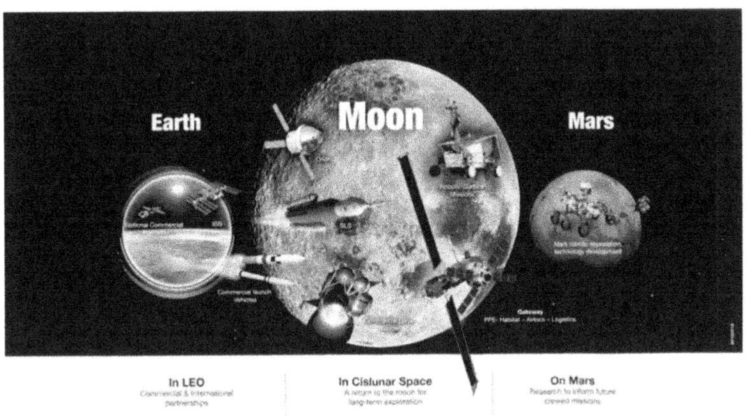

While that will be the first time the wheels of the 2,314-pound (1,050-kilogram) rover touch the Red Planet, the vehicle's network of processors, sensors and transmitters will, by then, have successfully simulated touchdown at Jezero many times before. Lonely Vigil: With the back shell that will help protect the Mars 2020 rover during its descent into the Martian atmosphere visible in the foreground, a technician on the project monitors the progress of Systems Test 1.

Full image and caption. "We first landed on Mars Jezero Crater on Jan. 23rd," said Heather Bottom, systems engineer for the Mars 2020 mission at the Jet Propulsion Laboratory in Pasadena, California. "And the rover successfully landed again on Mars two days later." Bottom was the test lead for Systems Test 1, or ST1, the Mars 2020 engineering team's first opportunity to take the major components of the Mars 2020 mission for a test drive.

"ST1 was a massive undertaking," said Bottom. "It was our first chance to exercise the flight software we will fly on 2020 with the actual spacecraft components that will be heading to Mars - and make sure they not only operate as expected, but also interact with each other as expected.

Figure 4. Mar's Rover is landing to the surface of Mars
Source: https://www.nasa.gov/sites/default/files/thumbnails/image/pia22812-nasa.jpg

Figure 5. Landing again by NASA's Lunar Plan to 2024

"The heritage for Mars 2020's software goes back to the Mars Exploration Rovers (Spirit and Opportunity) and the Curiosity rover that has been exploring Mars' Gale Crater since 2012. But 2020 is a different mission with a different rover, a different set f science instruments and a different destination on Mars. Its software has to be tailored accordingly.

On December 2, 2014, NASA's Advanced Human Exploration Systems and Operations Mission Director Jason Crusan and Deputy Associate Administrator for Programs James Reuthner [19]announced tentative support for the Boeing "Affordable Mars Mission Design" including radiation shielding, centrifugal artificial gravity, in-transit consumable resupply, and a lander which can return.[20] Reuthner suggested that if adequate funding was forthcoming, the proposed mission would be expected in the early 2030s.[21]

9.2.5. NASA'S Exploration on Venus, Asteroids, Jupiter and Saturn

9.2.5.1. NASA's Exploration on Venus

The *Magellan* spacecraft, also referred to as the Venus Radar Mapper, was a 1,035-kilogram (2,282 lb.) robotic space probe launched by NASA of the United States, on May 4, 1989, to map the surface of Venus by using synthetic aperture radar and to measure the planetary gravitational field. The *Magellan* probe was the first interplanetary mission to be launched from the Space Shuttle, the first one to use the Inertial Upper Stage booster for launching, and the first spacecraft to test aerobraking as a method for circularizing its orbit.

Magellan was the fifth successful NASA mission to Venus, and it ended an eleven-year gap in U.S. interplanetary probe launches.[22] Second planet from the Sun and our closest planetary neighbor, Venus is similar in structure and size to Earth, but it is now a very different world. Venus spins slowly in the opposite direction most planets do. Its thick atmosphere traps heat in a runaway greenhouse effect, making it the hottest planet in our solar system—with surface temperatures hot enough to melt lead. Glimpses below the clouds reveal volcanoes and deformed mountains.[23]

9.2.5.2. NASA's Exploration on the Asteroids

They are found in major asteroid belts and orbit between Mars and Jupiter. Our solar system's small bodies—asteroids, comets and meteors—pack big surprises. Asteroids and comets—and the meteors that sometimes come from them—are leftovers from the formation of our solar system 4.6 billion years ago.

While the planets and moons have changed over the millennia, many of these small chunks of ice, rock and metal have not. They are a lot like a fossil record of planetary evolution. There are currently 957,806 known asteroids and 3,620 known comets.[24] Two robotic spacecraft are currently exploring asteroids up close: NASA's OSIRIS-REx is in orbit at asteroid Bennu and Japan's Hayabusa 2 is preparing to collect samples from asteroid Ryugu.

Meanwhile, NASA's NEOWISE spacecraft, orbiting Earth, continues to improve on the most accurate survey of near-Earth objects every undertaken.

The Hubble Space Telescope and ground-based radar observatories also contribute regularly to our understanding of asteroids. Several more missions, including NASA's Psyche and Lucy, missions are in development to keep exploring these small worlds. Scientists also use ground-based radar to explore nearby asteroids whenever possible NASA's Galileo mission was the first spacecraft to fly past an asteroid.

Figure 6. Many Asteroid Belts in Space
Source: https://images.search.yahoo.com/search/images?p=NASA%27s+Exploration+on+

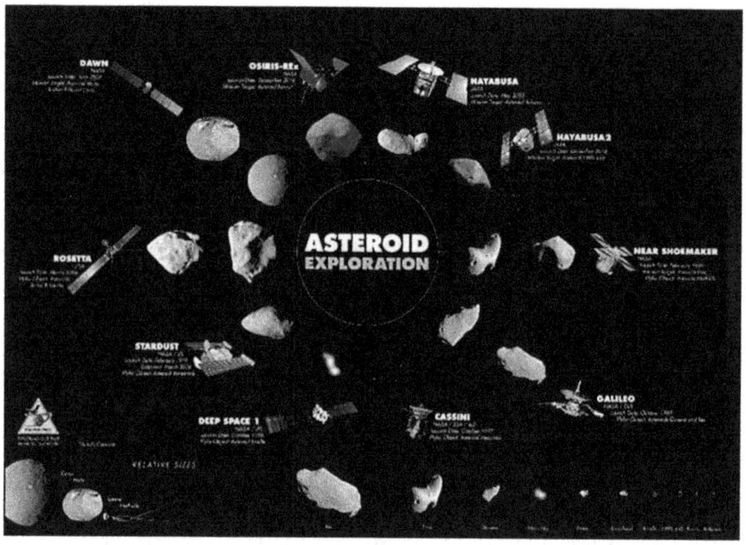

It flew past asteroid Gaspara in 1991 and Ida in 1993.

- NASA's Near-Earth Asteroid Rendezvous (NEAR-Shoemaker) mission studied asteroids Mathilde and Eros; and the Rosetta mission encountered Steins in 2008 and Lutetia in 2010. Deep Space 1 and Stardust both had close encounters with asteroids.
- ASA's Dawn spacecraft, launched in 2007, orbited and explored asteroid Vesta for over a year. Once it left in September 2012, it headed towards dwarf planet Ceres, with a planned arrival of 2015. Vesta and Ceres are two of the largest surviving protoplanetary bodies that almost became planets. By studying them with the same complement of instruments on board the same spacecraft, scientists will be able to compare and contrast the different evolutionary path each object took to help understand the early solar system overall. NASA's OSIRIS-REx will be the first U.S. mission to bring an asteroid sample to Earth.
- The spacecraft is currently orbiting asteroid Bennu and will spend two years mapping it before collecting a sample and returning to Earth.
- Bennu is a potentially hazardous asteroid that could one day threaten Earth.
- The sample will weigh at least 2.1 ounces (59.5 grams).
- The sample will be stored on the spacecraft and delivered to Earth in September 2023.[25]

9.2.5.3. NASA's Exploration on Jupiter

Jupiter has a long history surprising scientists—all the way back to 1610 when Galileo Galilei found the first moons beyond Earth. That discovery changed the way we see the universe. Fifth in line from the Sun, Jupiter is, by far, the largest planet in the solar system – more than twice as massive as all the other planets combined. Jupiter's familiar stripes and swirls are actually cold, windy clouds of ammonia and water, floating in an atmosphere of hydrogen and helium.

Jupiter's iconic Great Red Spot is a giant storm bigger than Earth that has raged for hundreds of years. One spacecraft — NASA's Juno orbiter — is currently exploring this giant world.

Jupiter's iconic Great Red Spot is a giant storm bigger than Earth that has raged for hundreds of years. Jupiter is surrounded by dozens of moons. Jupiter also has several rings, but unlike the famous rings of Saturn, Jupiter's rings are very faint and made of dust, not ice.[26] More recently, this planet has been visited by passing spacecraft, orbiters and probes.

Pioneer 10 and 11 and Voyager 1 and 2 were the first to fly by Jupiter in the 1970s, and since then we've sent Galileo to orbit the gas giant and drop a probe into its atmosphere. Cassini took detailed photos of Jupiter on its way to neighboring Saturn, as did New Horizons on its quest for Pluto and the Kuiper Belt. NASA's Juno spacecraft, which arrived in the Jovian system in July 2016, is currently studying the giant planet from orbit.[27]

- 1995-2003: The Galileo spacecraft drops a probe into Jupiter's atmosphere and conducts extended observations of Jupiter and its moons and rings.
- 2000: Cassini makes its closest approach to Jupiter at a distance of approximately 6.2 million miles (10 million kilometers), taking a highly detailed true color mosaic photo of the gas giant.
- 2007: Images taken by NASA's New Horizons spacecraft, on the way to Pluto, show new perspectives on Jupiter's atmospheric storms, the rings, volcanic Io, and icy Europa.
- 2009: On 20 July, almost exactly 15 years after fragments of comet Shoemaker-Levy slammed into Jupiter, a comet or asteroid crashes into the giant planet's southern hemisphere.
- 2011: Juno launches to examine Jupiter's chemistry, atmosphere, interior structure and magnetosphere.
- 2016: NASA's Juno spacecraft arrives at Jupiter, conducting an in-depth investigation of the planet's atmosphere, deep structure and magnetosphere for clues to its origin and evolution.[28]

9.2.5.4. NASA's Exploration on Saturn

Saturn is the sixth planet from the Sun and the second largest planet in our solar system. Adorned with thousands of beautiful ringlets, Saturn is unique among the planets. It is not the only planet to have rings—made of chunks of ice and rock—but none are as spectacular or as complicated as Saturn's. Like fellow gas giant Jupiter, Saturn is a massive ball made mostly of hydrogen and helium.[29]

With a radius of 36,183.7 miles (58,232 kilometers), Saturn is 9 times wider than Earth. If Earth were the size of a nickel, Saturn would be about as big as a volleyball.

From an average distance of 886 million miles (1.4 billion kilometers), Saturn is 9.5 astronomical units away from the Sun. One astronomical unit (abbreviated as AU), is the distance from the Sun to Earth. From this distance, it takes sunlight 80 minutes to travel from the Sun to Saturn.[30]

Four robotic spacecraft have visited Saturn. NASA's Pioneer 11 provided the first close look in September 1979. NASA's twin Voyager 1 and Voyager 2 spacecraft followed up with flybys nine months apart in 1980 and 1981. Each flyby revealed intriguing details about the ringed giant world, but it wasn't until the international Cassini mission arrived in orbit in 2004 that our understanding of Saturn really started to take shape.

Cassini studied Saturn from orbit for 13 years before its human engineers on Earth transformed it into an atmospheric probe for its spectacular final plunge into the planet in September 2017. Cassini also carried ESA's Huygens Probe, which landed on Saturn's moon Titan in 2005.

Figure 7. Saturn Image

Cassini mission ended a 13-year orbital mission with a spectacular, planned plunge into Saturn's atmosphere — sending science data back to the last second. Cassini's final five orbits enable scientists to directly sample Saturn's atmosphere for the first time on September 2017.[31]

9.2.5. NASA

The National Aeronautics and Space Administration (NASA), is an independent agency of the United States Federal Government responsible for the civilian space program, as well as aeronautics and aerospace research. NASA was established in 1958, succeeding the National Advisory Committee for Aeronautics (NACA). The new agency was to have a distinctly civilian orientation, encouraging peaceful applications in space science.[32]

Since its establishment, most US space exploration efforts have been led by NASA, including the Apollo Moon landing missions, the Skylab space station, and later the Space Shuttle.

NASA is supporting the International Space Station and is overseeing the development of the Orion Multi-Purpose Crew Vehicle, the Space Launch System and Commercial Crew vehicles. The agency is also responsible for the Launch Services Program which provides oversight of launch operations and countdown management for unscrewed NASA launches.[33]

NASA science is focused on better understanding Earth through the Earth Observing System; advancing heliophysics through the efforts of the Science Mission Directorate's Heliophysics Research Program; exploring bodies throughout the Solar System with advanced robotic spacecraft missions such as *New Horizons*;[34] and researching astrophysics topics, such as the Big Bang, through the Great Observatories and associated programs. US law requires the International System of Units to be used in

Figure 8. NASA Organization Chart
Source: https://www.researchgate.net/figure/1-NASA-Organization-Chart-The-johnsoon-Space-Center-was-established-in-1961-as-the-Manned_fig1_236962693

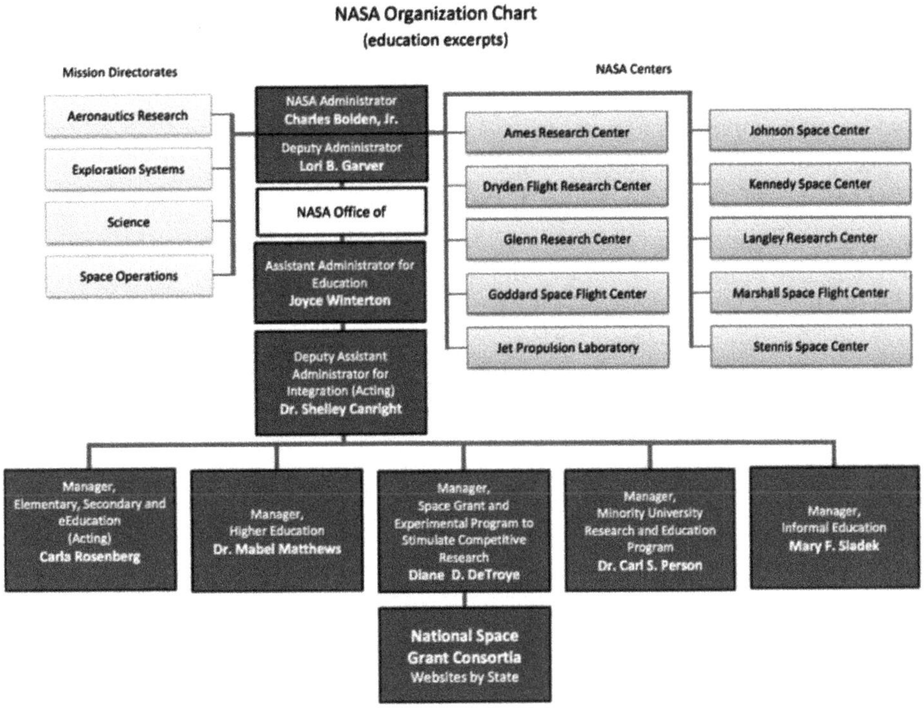

all U.S. Government programs, "except where impractical".[35] Today NASA is predominantly working with SI units, but some projects still use a mix of US and SI units.

The NASA Authorization Act of 2017, which included $19.5 billion in funding for that fiscal year, directed NASA to get humans near or on the surface of Mars by the early 2030s. Though the agency is independent, the survival or discontinuation of projects can depend directly on the will of the President.[36]

NASA's next major space initiative is to be the construction of the Lunar Gateway. This initiative is to involve the construction of a new space station, which will have many features in common with the current International Space Station, except that it will be in orbit about the Moon, instead of the Earth.[37] space station will be designed primarily for non-continuous human habitation. The first tentative steps of returning to crewed lunar missions will be Artemis 2, which is to include the Orion crew module, propelled by the Space Launch System (SLS)[38], and is to launch in 2022 into a Lunar flyby. The construction of the Gateway would begin with the proposed Artemis 3, which is planned to deliver a crew of four to Lunar orbit along with the first modules of the Gateway.

This mission would last for up to 30 days. NASA plans to build full scale deep space habitats such as the Lunar Gateway[39] and the Nautilus-X as part of its Next Space Technologies for Exploration Partnerships (Next STEP) program. In 2017, NASA was directed by the congressional NASA Transition Authorization Act of 2017 to get humans to Mars-orbit (or to the Martian surface) by 2030s.[40]

ENDNOTES

1 https://en.wikipedia.org/wiki/Canadian_Space_Agency

2 "*ESA and Canada renew partnership in space science and technology*". European Space Agency, 15 December 2010. Retrieved 2014-02-02.

3 https://spacepolicyonline.com/topics/space-law/

4 https://spacepolicyonline.com/topics/space-law

5 https://spacepolicyonline.com/topics/space-law/

6 Jackson, Shanessa (11 September 2018). *"Competition Seeks University Concepts for Gateway and Deep Space Exploration Capabilities"*. nasa.gov. NASA. Retrieved 19 September 2018.

7 Gebhardt, Chris (6 April 2017). *"NASA finally sets goals, missions for SLS – eyes multi-step plan to Mars"*. NASASpaceflight.com. NASA Spaceflight. Retrieved 19 September 2018.

8 Mahoney, Erin (24 August 2018). *"NASA Seeks Ideas for Scientific Activities Near the Moon"*. nasa.gov. NASA. Retrieved 19 September 2018.

9 Weitering, Hanneke (27 September 2017). *"NASA and Russia Partner Up for Crewed Deep-Space Missions"*. Space.com. Retrieved 5 November 2017.

10 NASA (2 May 2018). *"Gateway Memorandum for the Record"* (PDF). nasa.gov. NASA. Retrieved 19 September 2018.

11 https://www.nasa.gov/sites/default/files/atoms/files/fy_2021_budget_book_508.pdf-p.4|access-date=2020-04-01

12 "NASA just got its best budget in a decade". www.planetary.org. Retrieved 27 February 2019.

13 Foust, Jeff (23 July 2019). *"NASA to sole source Gateway habitation module to Northrop Grumman"*. SpaceNews. Retrieved 11 December 2019.

14 Halo orbit selected for Gateway space station. David Szondy, *New Atlas*. 18 July 2019.

15 https://www.theverge.com/2019/6/7/18656839/trump-nasa-moon-mars-tweet-artemis-lunar-exploration

16 https://spacenews.com/nasa-outlines-plan-for-2024-lunar-landing

17 https://www.nasa.gov/specials/moon2mars

18 Hubbard, G. Scott (2012-08-28). *"A Next Decade Mars Program"*. The Huffington Post. Retrieved 18 October 2012.

19 https://en.wikipedia.org/wiki/Exploration_of_Mars

20 K.Klaus, M. L. Raftery and K. E. Post (2014) *"An Affordable Mars Mission Design"* Archived 2015-05-07 at the Wayback Machine (Houston, Texas: Boeing Co.)

21 NASA (December 2, 2014) *"NASA's Journey to Mars News Briefing" NASA TV*

22 https://en.wikipedia.org/wiki/Magellan_(spacecraft)

23 https://solarsystem.nasa.gov/planets/venus/overview

24 https://solarsystem.nasa.gov/asteroids-comets-and-meteors/in-depth

25 https://solarsystem.nasa.gov/missions/osiris-rex/in-depth

26 https://solarsystem.nasa.gov/planets/jupiter/in-depth/

27 https://solarsystem.nasa.gov/planets/jupiter/exploration/?page=0&per_page=10&order=launch_date+desc%2Ctitle+asc&search=&tags=Jupiter&category=33

28 https://solarsystem.nasa.gov/planets/jupiter/exploration/?page=0&per_page=10&order=launch_date+desc%2Ctitle+asc&search=&tags=Jupiter&category=33

29 https://solarsystem.nasa.gov/planets/saturn/overview

30 https://solarsystem.nasa.gov/planets/saturn/in-depth

31 https://solarsystem.nasa.gov/planets/saturn/exploration/?page=0&per_page=10&order=launch_date+desc%2Ctitle+asc&search=&tags=Saturn&category=33

32 *"Ike in History: Eisenhower Creates NASA"*. Eisenhower Memorial. 2013. Archived from the original on November 19, 2013. Retrieved November 27, 2013.

33 https://en.wikipedia.org/wiki/NASA

34 Roston, Michael (August 28, 2015). *"NASA's Next Horizon in Space"*. New York Times. Archived from the original on August 29, 2015. Retrieved August 28, 2015.

35 nternational System of Units - The Metric Measurement System | NASA

36 Fouriezos, Nick (May 30, 2016). "Your Presidential Candidates ... For the Milky Way". OZY. Archived from the original on May 30, 2016. Retrieved May 30, 2016.

37 "*NASA Sets New Roadmap for Moon Base, Crewed Missions to Mars*" archived on November 27, 2018, at the Wayback Machine Extreme Tech. By Ryan Whitwam. September 27, 2018. Downloaded November 26, 2018.

38 https://en.wikipedia.org/wiki/Space_Launch_System

39 https://en.wikipedia.org/wiki/Lunar_Gateway

40 https://thehill.com/opinion/technology/362356-lets-get-humans-to-mars-in-the-2030s

Chapter 10
South America's National Legislation on the Outer Space Laws and Organizations

ABSTRACT

This chapter talks about the different countries in South America and their laws around outer space.

10.1. ARGENTINA

The Argentina has two decrees on outer space actives: "*Establishment of the National Registry of Objects Launched into Outer Space*" National Decree No.125 of 25 July 1995 (original version). "*Creation of the National Commission on Space Activities* (CONAE)" National Decree No. 995 of 28 May 1991.[1]

As the title suggests, the former was enacted to establish a national registry of space objects along with registration procedure so as to comply with the 1975 Registration Convention.

The latter decree created the Argentina National Commission on Space Activities with the competence take care of formulate, implement, monitor, manage, and administer projects and undertakings in the area of space. Argentina's first activities in the space field go back to 1961, when the National Commission for Space Research (Comisión Nacional de Investigaciones Espaciales, CNIE) was first established within the Argentine Air Forces area. CNIE, working with local and international partners, carried out the first southern hemisphere scientific atmospheric physics using rockets and stratospheric balloons.

Together with the Argentine Institute of Aeronautics and Space Research, CNIE designed and constructed a family of one- and two-stage sounding rockets, i.e. the Orion, the Rigel and the Castor, which were launched from Chamical, in the Province of La Rioja.[2]

DOI: 10.4018/978-1-7998-7407-2.ch010

10.2. BRAZIL

The Brazilian legal framework applicable to space-related activities has been conceived in accordance with fundamental guidelines and rules instituted by the Federal Constitution of 1988. Such a framework represents a historical milestone as far as democratic movements in Latin America are concerned (Moraes, 2019), and it has played a major role in the advancement of the Brazilian national space program.

To improve the coordination of the Brazilian space program and to stress its peaceful purposes, the Brazilian Space Agency (Agência Espacial Brasileira, AEB) was created by Law 8.854/1994. As a civilian entity headquartered in Brasilia, it enjoys financial and administrative autonomy (Article 2).

Originally submitted to the Presidency of the Republic, by Decree no. 3.566/2000, the AEB has been under governance of the Ministry of Science, Technology, and Innovation (MCTI). According to its current regimental structure, the AEB comprises a Presidency, a Superior Council, and four dedicated Directories (Decree no. 10.469/ 2020).[3]Let's look at the Brazilian space program. It all started with President Jânio Quadros who established in the 1960s a commission that elaborated a national program for the space exploration. Then, the main objectives of the Brazilian space program were set in 1979 by the Brazilian Complete Space Mission. This initial program has been gradually extended.

The return to a civilian-led government after twenty years of military rule (between 1964 and 1985) as well as pressure from the United States of America for the space program to be demilitarized, led in 1994 to the creation of the Brazilian Space Agency (AEB) which replaced the inter-ministerial structure which depended on the military. Recently, in 2011, Argentina's Minister of Defense, Arturo Puricelli, made a proposal to Celso Amorim for the creation of a unified South American space agency by the year 2025.[4]

The Brazilian Space Agency was established by "*Law of Establishing the Brazilian Space Agency.*" The Brazilian Space Agency (Portuguese: *Agência Espacial Brasileira*; AEB) is the civilian authority in Brazil responsible for the country's space program.

It operates a spaceport at Alcântara, and a rocket launch site at Barreira do Inferno. The agency has given Brazil a role in space in South America and made Brazil a former partner for cooperation in the International Space Station.[5]

The Brazilian Space Agency is the heir to Brazil's space program. Previously, the program had been under the control of the Brazilian military; the program was transferred into civilian control on 10 February 1994.

It suffered a major setback in 2003, when a rocket explosion killed 21 technicians. Brazil successfully launched its first rocket into space on 23 October 2004 from the Alcântara Launch Center; it was a VSB-30 launched on a sub-orbital mission. Several other successful launches have followed.[6]

On March 30, 2006, AEB astronaut Marcos Pontes became the first Brazilian and the first native Portuguese-speaking person to go into space, where he stayed on the International Space Station for a week.

During his trip, Pontes carried out eight experiments selected by the Brazilian Space Agency. These experiments included testing flight dynamics of saw blades in zero gravity environments, and attempting to "tip over" a table of space food".

He landed in Kazakhstan on April 8, 2006, with the crew of Expedition 12.[7]

The Brazilian Space Agency has pursued a policy of joint technological development with more advanced space programs. Initially, it relied heavily on the United States, but after meeting difficulties from them on technological transfers, Brazil has branched out, working with other nations, including China, India, Russia, and Ukraine[REMOVED TA FIELD].

10.3. CHILE

"Decree concerning the Establishment of a Presidential Advisory Committee known as the Chilean Space Agency" was promulgated by Supreme Decree No. 338 of 17 July 2001. A Presidential Advisory Committee known as the Chilean Space Agency shall be established to provide advice in all matters concerning the identification, formulation and implementation of policies, plans, programmes, measures and other activities relating to space, and to serve as coordinating center government.

Chile joined the race for space in 2001 by launching its first-ever space agency. The Chilean Space Agency was to coordinate scientific research as well as provide a focus for companies interested in the commercial possibilities of space technology. In particular, the government wanted the agency to encourage the development of a Chilean satellite construction program. [8]

There are still a lot of decisions and a lot of work ahead. But the important thing is that, for the first time, we now have a national space agency that can begin a dialogue with other international agencies." CNES and the Chilean Space Agency sign framework cooperation agreement Press Release From: Centre National d'Etudes Spatiales (CNES) Posted: Thursday, April 4, 2002 . Mr. Alain Bensoussan, President of the Centre National d'Etudes Spatiales (CNES), and Nelson Hadad Heresy, President of the Chilean Space Agency (ACE), signed a framework agreement governing cooperation in space technologies and applications in Santiago, Chile, April, 2002. This agreement, which will run for a renewable term of four years, stems from the two agencies' desire to work together in space, particularly for the purposes of environmental monitoring. Earth observation and space technology development. The agreement also aims to foster mutually beneficial industrial cooperation initiatives in space between the two nations, and to exploit space applications to boost economic, social and scientific development.[9]

ENDNOTES

1. http://www.esa.int/About_Us/ECSL_European_Centre_for_Space_Law/National_Space_Laws
2. https://spacegeneration.org/regions/south-america/argentina
3. https://oxfordre.com/planetaryscience/view/10.1093/acrefore/9780190647926.001.0001/acrefore-9780190647926-e-213
4. https://www.spacelegalissues.com/space-law-the-birth-of-the-brazilian-space-program
5. Brazilian Space Agency Archived 2015-09-23 at the Wayback Machine The Internet Encyclopedia of Science. Retrieved on 2009-07-29.
6. Herald Tribune: Brazil launches rocket for gravity research International Herald Tribune. Retrieved on 2009-07-30.
7. BBC World: First Brazilian goes into space BBC News. Retrieved on 2009-07-30.
8. https://www.globalsecurity.org/space/world/chile/agency.htm
9. http://www.spaceref.com/news/viewpr.html?pid=7914

Chapter 11
Russian Federation, Kazakhstan, and Ukraine's National Legislation on the Outer Space Laws and Organizations

ABSTRACT

This chapter will speak to the laws regarding outer space and other organizations in countries such as Russia, Kazakhstan, and the Ukraine.

11.1. RUSSIAN FEDERATION

11.1.1. Russian Space Activity Law

The "Law of the Russian Federation (No. 5663-1 of August 20, 1993) on Space Activities" was amended and add ended with November 29, 1996, January 10, 2003, March 5, August 22, 2004 and February 2, December 18, 2006.[1]

This Law was composed of thirty Articles and seven Sections as the following;

Section I. General Provisions
Article 2. The Meaning of Space Activity
Article 4. Principles of Space Activity
Section II. Organization of Space Activity
Article 5. Supervision of Space Activity
Article 8, The Federal Space Program of Russia
Article 9. The Licensing of Space Activity

DOI: 10.4018/978-1-7998-7407-2.ch011

Article 10. Certification of Space Technology
Section III. Economic Conditions of Space Activity
Article 12. Financing Space Activity
Section IV. Space Facilities. Space Infrastructure
Article 17. Space Projects
Article 18. Space Infrastructure
Article 20. Cosmonauts and Crews of Manned Space Objects
Section V. Safety of Space Activities
Article 22. Ensuring the Safety of Space Activities
Article 23. Investigation of Incidents during the Implementation of Space Activities
Article 25. Insurance of Space Activity
Section VI. International Cooperation
Article 28. Legal Regulation of International Cooperation
Section VII. Liability
Article 29. Liability of Officials, Organizations, and Citizens
Article 30. Liability for Harm Caused When Implementing space Activity

11.1.2. Russia: Luna Program 25, 26, and 27; July 2021, 2024, and 2025

The last mission under the Luna program was in 1976, when the world still had a Soviet Union. Eager to make sure it's got a stake in the 21st-century moon rush, Russia is resurrecting the program with a slate of new missions focused on enabling future mining operations.

Luna 25 will prove out new landing technologies and drill into the surface of the south pole to study the composition of the moon's soil for water ice, helium-3, carbon and nitrogen, precious metals, and any other interesting resources.

Luna 26 is an orbiter that will survey the landscape from above to try to get a sense of where these resources might lie. And Luna 27 is another lander headed to the south pole to specifically prospect for water ice.[2]Luna 25 (Luna-Glob lander)[3] is a planned lunar lander mission by the Russian Federal Space Agency (Roscosmos). It will land near the lunar south pole at the Boguslavsky crater. It was renamed from Luna-Glob lander to Luna 25 to emphasize the continuity of the Soviet Luna Programme from the 1970s, though it is still part of what was at one point conceptualized as the Luna-Glob lunar exploration program. The launch is scheduled for October 2021.

Luna 27 (Luna-Rescues lander) is a planned lunar lander mission by the Russian Federal Space Agency (Roscosmos) with collaboration by the European Space Agency (ESA) to send a lander to the South Pole–Aitken basin, an area on the far side of the Moon. Its objective will be to detect and characterize lunar polar volatiles. The mission is a continuation of the Luna-Glob programme. Luna 27 (Luna-Resurs lander)[4] is a planned lunar lander mission by the Russian Federal Space Agency (Roscosmos) with collaboration by the European Space Agency (ESA) to send a lander to the South Pole–Aitken basin, an area on the far side of the Moon. Its objective will be to detect and characterize lunar polar volatiles. The mission is a continuation of the Luna-Glob programme.

The purpose is to prospect for minerals, volatiles (nitrogen, water, carbon dioxide, ammonia, hydrogen, methane and sulfur dioxide.), and lunar water ice in permanently shadowed areas of the Moon and investigate the potential use of these natural lunar resources.[5] On the long term, Russia considers building a crewed base on the Moon's far side that would bring scientific and commercial benefit.

11.1.3. Russia Plans to Send Cosmonauts to the Moon and to Mars, Venus and Jupiter etc. by 2030

Russia plans to send cosmonauts to the moon and unmanned spacecraft to Mars, Venus and Jupiter, all by 2030. These ambitious spaceflight goals are laid out in a strategy document drawn up recently by Russia's Federal Space Agency (known as Roscosmos). And there's more. Roscosmos wants a new rocket called Angara to become the nation's workhorse launch vehicle by 2020, replacing the venerable Soyuz and Proton rockets that have been carrying the load since the 1960s. The space agency also plans to top Angara with a new six-seat spaceship, an upgrade over the three-passenger Soyuz spacecraft that is currently the world's only means of transporting astronauts to and from the International Space Station. Angara will launch from a new spaceport in eastern Russia called Vostochny, Russia's new space vision focuses heavily on the moon. In addition to the manned lunar landing, Roscosmos is considering building a space station in orbit around Earth's nearest neighbor by 2030. The Russian space plan also calls for sending robotic probes to visit Venus, Jupiter and Mars by 2030.[6] Roscosmos' goals may strike some observers as incredibly ambitious.

11.1. 4. The Roscosmos State Corporation for Space Activities

The Roscosmos State Corporation for Space Activities (commonly known as Roscosmos (Russian: Роскосмос) is a state corporation responsible for the wide range and types of space flights and cosmonautics programs for the Russian Federation.[7]

Figure 1. Roscosmos Building located in Moscow, Soyuz Sat. & Space State
Source: https://en.wikipedia.org/wiki/Roscosmos

The launch facilities used are Baikonur Cosmodrome in Kazakhstan (with most launches taking place there, both crewed and uncrewed), and Vostochny Cosmodrome being built in the Russian Far East in Amur Oblast. The current director since May 2018 is Dmitry Rogozin. In 2015 the Russian government merged Roscosmos with the United Rocket and Space Corporation, the re-nationalized Russian space industry, to create the Roscosmos State Corporation.[8] The Hall of Space Technology in the Tsiolkovsky State Museum of the history of Cosmonautics, Kaluga, Russia.[9]

11.2. KAZAKHSTAN

11.2.1. The Kazakhstan Space Activity Law

The Law of the Republic of Kazakhstan on Space Activity was enacted on January 6, 2012 (No. 528-IV). This Law regulates public relations on carrying out of space activity in the Republic of Kazakhstan.[10]

The Space Activity Law of Kazakhstan consisted of thirty seven articles and seven Chapters as the following;

Chapter 1. GENERAL PROVISIONS
Article 1. Basic definitions, used in this Law
Article 3. Principles of carrying out the space activity
Chapter 2. STATE REGULATION AND CONTROL IN THE FIELD OF SPACE ACTIV
Article 8. Competence of the Government of the Republic of Kazakhstan in the field of space activity shall be prohibited
Article 11. State registration of the space objects and rights to them
Chapter 3. CARRYING OUT OF SPACE ACTIVITY
Article 13. Licensure of the activity in the scope of use of cosmic space
Article 14. Scientific researches in the field of space activity
Article 15. Creation of space systems and space missile weapon systems
Article 17. Use of space system of the Earth's remote sensing
Chapter 4. OUTER SPACE INFRASTRUCTURE
Article 20. Objects of outer space infrastructure of the Republic of Kazakhstan
Article 21. Objects of ground outer space infrastructure
Article 23. "Baikonur" launch area
Chapter 5. SAFETY OF THE SPACE ACTIVITY
Article 27. Promotion of safety of the space activity
Article 28. Investigation of accidents upon carrying out of the space activity
Article 30. Prohibitions and restrictions in the space activity
Chapter 6. LEGAL STATUS AND MEASURES OF SOCIAL SECURITY OF THE ASTRONAUT CANDIDATE, ASTRONAUT
Article 31. Status of astronaut candidate, astronaut. Preparation of the astronaut candidate, astronaut
Article 32. Guarantees in case of receiving the maim, disease or death incident (death) of the astronaut candidate, astronaut
Article 34. Medical and sanitary and sanatorium-resort service of the astronaut candidate, astronaut

Article 35. Financial security of astronauts, honored by honorary title "Қазақстанның ғарышкер-ұшқышы" (Pilot-astronaut of Kazakhstan)

Chapter 7. FINAL PROVISIONS

Article 36. Responsibility for violation of the legislation of the Republic of Kazakhstan in the field of space activity Violation of the legislation of the Republic of Kazakhstan in the field of space activity entailed responsibility in accordance with the Laws of the Republic of Kazakhstan

Article 37. The order of enforcement of this Law, This Law enters into force upon expiry of ten calendar days after its first official publication.

11.2.2. Kazakhstan National Space Agency

The National Space Agency of the Republic of Kazakhstan is Kazakhstan's national space agency, and was officially established on March 27, 2007. On 7 January 2000, the Kazakh government decreed it would form a cosmonaut corps. [11]

KazCosmos signed a contract with ISS-Reshetnev and Thales Alenia Space Italy on June 21, 2011 for the third telecommunications satellite, named KazSat-3 and launched it in 2014.

Figure 2. World's Oldest Space Launch Facility: The Baikonur Cosmodrome
Source: https://en.wikipedia.org/wiki/Baikonur_Cosmodrome

Aidyn Aimbetov launched from Baikonaur on 2 September 2015 aboard Soyuz TMA-18M, docked with the International Space Station, and returned on 11 September 2015 aboard Soyuz TMA-16M, touching down on the Kazakhstan Steppe, having spent 10 days in space.[12]

The Baikonur Cosmodrome (Kazakh) is a spaceport located in an area of southern Kazakhstan leased to Russia. The Cosmodrome is the world's first and largest operational space launch facility.[13] The spaceport is located in the desert steppe of Baikonur, about 200 kilometers (124 mi) east of the Aral Sea and north of the river Syr Darya.

It is near the Tyuratam railway station and is about 90 meters (300 ft.) above sea level. Baikonur Cosmodrome and the city of Baikonur celebrated the 63rd anniversary of the foundation on 2 June 2018. The spaceport is currently leased by the Kazakh Government to Russia until 2050, and is managed jointly by the Roscosmos State Corporation and the Russian Aerospace Forces.

11.3. UKRAINE

11.3.1. New Space Activities Law

The Ukrainian parliament passed the country's first commercial space law, titled "*On Amendments to Certain Laws of Ukraine on the State Regulation of Space Activities*," that was signed on 25 October 2019 by Ukrainian President Volodymyr Zelensky, according to Oleksiy Burchevskyy of the Kinstellar law firm writing in *Lexology*.

Until this legislation all space activities in Ukraine have been carried out solely by state operators. As of 29 January 2020, most of the new commercial space law will come into force, with the remaining provisions of the legislation coming into effect by the end of October 2020.

The new Ukrainian commercial space law will allow private Ukrainian citizens and companies to commercially exploit and explore Earth orbit and beyond from Ukraine and under Ukrainian jurisdiction. A parallel aim, apparently, is to also reduce the role in the supervision of Ukrainian space activities by traditional state operators. Under the new law, private companies seeking to conduct commercial space activities in and from Ukraine will have to submit a declaration to government authorities. Commercial space activities that will involve rocket engine testing, satellite launch, and the control of satellites from Ukrainian territory will require a state permit.[14]

The importance of promoting international cooperation, maintaining and developing existing international relations in the space industry, taking into account national interests, is also emphasized in the law. Currently, only state-owned companies can carry out activities related to testing, production and operation of space launch vehicles Ukraine.[15] The Ordinance Of The Supreme Soviet Of Ukraine, On Space Activity Law of Ukraine had enacted on 15 November 1996 (VVRU, 1997, p. 2).

This Law composed of 29 Articles and seven Sections as the following;

Section 1. General Provisions (Art. 1~4),
Section 2. Organization of Space Activity (Art. 5~11),
Section 3. General Requirements Imposed on Space Facilities (Art. 12~16)
Section 4. Participation by Ukraine in International Space Related Cooperation (Art.17~ 19)
Section 5. Ensuring the Safety of Space Activity (Art. 20~25)
Section 6. Space Activity Related to the Defense and Security of Ukraine (Art. 26~28)
Section 7. Final Provisions (Art.29)

The importance of promoting international cooperation, maintaining and developing existing international relations in the space industry, taking into account national interests, is also emphasized in the law.

Under the new law, state-owned companies will not be required to apply for licenses to work in the space sector. Currently, only state-owned companies can carry out activities related to testing, production and operation of space launch vehicles in Ukraine.

11.3.2. The State Space Agency of Ukraine

The State Space Agency of Ukraine (SSAU) was established by the Decree of the President of Ukraine on the establishment of the National Space Agency of Ukraine (No. 117) on February 29, 1992. This

Decree (No. 665/97) was amended on July 22, 1997. The State Space Agency of Ukraine is the Ukrainian government agency responsible for space policy and programs.

The State Space Agency of Ukraine does not specialize in manned Astronautical programs.

It is the second of two direct Soviet space program descendants. The agency does not have its own spaceport and until 2014, depended on the resources of the Russian Federal Space Agency (the primary inheritor of the Soviet space program). [16]Until December 9, 2010, the agency was known as the National Space Agency of Ukraine (NSAU).[17]·Until 2014 launches were conducted at Kazakhstan's Baikonur and Russia's Plesetsk Cosmodromes. After the Russian annexation of Crimea, launches were conducted on Sea Launch's floating platform, which was soon mothballed. NSAU has ground control and tracking facilities in Kiev and a control center in Dunaivtsi (Khmelnytskyi Oblast). facilities in Yevpatoria, Crimea were abandoned after the annexation by Russia. Ukrainian spacecraft include a few kinds for domestic and foreign use and international cooperation.

ENDNOTES

1 https://www.wto.org/english/thewto_e/acc_e/rus_e/WTACCRUS58_LEG_375.pdf

2 https://en.wikipedia.org/wiki/Luna_27

3 "*Missions to the Moon*". The Planetary Society. Retrieved 14 January 2020.

4 https://en.wikipedia.org/wiki/Luna_27

5 Ghosh, Pallab (16 October 2015). *"Europe and Russia mission to assess Moon settlement"*. BBC News. Retrieved 2015-10-16.

6 https://www.space.com/14915-russia-moonmoon-landing-2030.html

7 https://en.wikipedia.org/wiki/Roscosmos

8 "Russia Merges United Rocket and Space Corporation with Roscosmos". Via Satellite. 23 January 2015. Retrieved 10 February 2015.

9 https://en.wikipedia.org/wiki/Roscosmos

10 http://bayterek.kz/en/info/zakon%20o%20kosmose.php

11 https://en.wikipedia.org/wiki/KazCosmos

12 Irene Klotz (11 September 2015). *"Record-setting cosmonaut, two visiting crewmen head home from space station"*. Reuters. Science Daily. Archived from the original on 14 September 2015.

13 https://en.wikipedia.org/wiki/Baikonur_Cosmodrome

14 https://spacewatch.global/2019/11/ukraine-passes-commercial-space-law-allowing-private-space-in-2020

15 https://open4business.com.ua/ukraines-parliament-opens-space-for-private-companies/

16 https://en.wikipedia.org/wiki/State_Space_Agency_of_Ukraine[REMOVEDTAFIELD]

17 Указ Президента України № 1085 від 9 грудня 2010 року «Про оптимізацію системи центральних органів виконавчої влади» Archived 2012-02-10 at the Wayback Machine *(Ukrainian)*

Chapter 12
Australia's National Legislation on the Outer Space Laws and Organizations

ABSTRACT

This chapter discusses outer space laws and national legislation in outer space.

12.1. AUSTRALIA

12.1.1. Australian Space Activities Act

In Australia, the launching of rockets and activities associated with getting satellites into space have been regulated at the Federal Government level since 1998. The Space Activities Act was introduced following proposals by a number of companies to open commercial space ports across the country in the mid-1990s.

The *Space Activities Act* of 1998, *Space Activities Regulations* of 2001 (Regulations), and the *Space Activities (Scientific or Educational Organizations) Guidelines* of 2015 create a regulatory framework for civilian space activities in Australia, as well as for those conducted overseas but involving Australian interests. The Act and Regulations deal primarily with the launching of space objects and the return to Australia of space objects.

The Australian government is undertaking a review of the *"appropriateness and effectiveness"* of the Space Activities Act of 1998. On 24 October 2015, the Minister for Industry, Innovation and Science of Australia, the Hon Christopher Pyne MP, announced that the Government will conduct a review of the Space Activities Act 1998 and Space Activities Regulations 2001. The regulatory framework for Australian civil space activities is complex and must consider diverse and often conflicting issues. The department is inviting stakeholders to provide submissions to the review via an online survey from 24 February to 30 April 2016.

DOI: 10.4018/978-1-7998-7407-2.ch012

The Australian Space Activities Act of 1998 composed of eight parts;

Part 1—Introduction
Part 2—Definitions
Part 3—Regulation of space activities
Part 4—Liability for damage by space objects
Part 5—Register of space objects
Part 5A—Implementation of space cooperation agreements
Part 6—Civil penalties
Part 7—Investigation of accidents
Part 8—Miscellaneous
Schedule 1—Convention on International Liability for Damage Caused by Space Objects
Schedule 2—Convention on Registration of Objects Launched into Outer Space
Schedule 3—Treaty on Principles Governing the Activities of States in the Exploration and Use of Outer Space, including the Moon and other Celestial Bodies
Schedule 4—Agreement Governing the Activities of States on the Moon and other Celestial Bodies
Schedule 5—Agreement on the Rescue of Astronauts, the Return of Astronauts and the Return of Objects Launched into Outer Space
Schedule 6—Agreement between the Government of Australia and the Government the Russian Federation Cooperation in the Field of the Exploration and Use of Outer Space for Peaceful Purposes.

12.1.2. ILA 78th 2018 Sydney International Space Law Conference

Australia was played a key role in the continuing development of space law at the 78th Biennial (ILA) Sydney International Law Conference.

This 78th Biennial Conference of the International Law Association (ILA) hosted by ILA Australia Branch was held at the InterContinental Hotel in Sydney, Australia on August 19~24, 2018.

As I am now a member of the ILA Space Law Committee for 22 years since ILA 67th Helsinki Conference of 1996, so I presented with my title entitled *"Proposal for Establishing a New International Space Organization on the Moon & Mars Exploitation and Legal Problems on the Moon Agreement"* at the Space Law Committee of the 78th I LA International Law Conference in Sydney, Australia on August 22, 2018.

A subset of the ILA, the Space Law Committee, was debated legal developments in the use and exploration of space.

It was also considered the international regulation of space activities, including their impact on Australia and the South-East Asian region.

Australia has had limited involvement with the UN space agencies and similar international organizations, such as the Paris-based International Institute of Space Law, formed to manage the use of space.

Australia's commitment to the commercial use of space, demonstrated by the establishment of the ASA and associated project funding, confers a growing responsibility to contribute to the international debate concerning the fundamental principles relating to space law.[1]

12.I.3. Australian Space Agency

The Australian Space Agency is Australia's national agency responsible for the development of Australia's commercial space industry, coordinating domestic activities, identifying opportunities and facilitating international space engagement that include Australian stakeholders. Its headquarters are located in Adelaide, the southeastern capital city of South Australia.[2]

The Agency has six primary responsibilities:[3]

1. Providing national policy and strategic advice on the civil space sector.
2. Coordinating Australia's domestic civil space sector activities.
3. Supporting the growth of Australia's space industry and the use of space across the broader economy.
4. Leading international civil space engagement.
5. Administering space activities legislation and delivering on our international obligations.
6. Inspiring the Australian community and the next generation of space entrepreneurs.

In keeping with the above responsibilities, it is different from other established space programs in that it exists to promote private development and businesses rather than state space operations (contrasting with NASA and the European Space Agency).

Australian scientists and researchers have, for some years, been heavily involved in new and emerging science and technology related to the exploration of outer space, satellite communications, weather and global positioning systems and related cyber technologies such as digitisation, drone technology and ground-based telemetry-based tracking.

As the Australian space industry grows and matures under the sponsorship of the new Australian Space agency (ASA), it will become increasingly important to engage with, and contribute towards, the international debate concerning space law and the priorities and principles governing outer space.[4]

On 12 December 2018, Prime Minister Scott Morrison officially announced that Adelaide would become home to the Australian Space Agency. It would be located at Lot Fourteen, the site of the former Royal Adelaide Hospital in 2019. The offices were officially opened on 19 February 2020. The Agency is planning to "triple the size of the Australian space industry and create 20,000 new jobs by 2030".[5]

ENDNOTES

1 https://www.futuredirections.org.au/publication/space-law-australian-opportunities-responsibilities

2 https://en.wikipedia.org/wiki/Australian_Space_Agency

3 "*Australian Space Agency Charter*" (PDF). Department of Industry, Innovation & Science. Australian Government. Retrieved 8 January 2019.

4 https://www.futuredirections.org.au/publication/space-law-australian-opportunities-responsibilities/

5 *"Australian Space Agency opens in Adelaide"*. Australia. Department of Industry, Science, Energy and Resources. 19 February 2020. Retrieved 19 February 2020.

Chapter 13
Africa's National Legislation on the Outer Space Laws and Organizations

ABSTRACT

This chapter discusses Africa's national legislation on outer space laws.

13.1. SOUTH AFRICA

13.1.1. Space Affairs Act

The Space Affairs Act (Statutes of the Republic of South Africa - Trade and Industry No. 84 of 1993) was assented on June 23, 1993 and then commenced on September, 6 1993.[1] This Space Affairs Act composed of twenty seven Articles as the following main key points.

Article 1. Definitions,
Article 4. Establishment of South African Council for Space Affairs,
Article 5. Objects and functions of Council,
Article 6. Composition of Council,
Article 7. Meetings of Council,
Article 9. Boards of inquiry,
Article 10. Inspectors,
Article 11. Licensing,
Article 14. Duties and liabilities of licensee,
Article 15. Actions in case of accident, incident or potential emergency,
Article 16. Appeal to Minister,
Article 17. Revision by court of law,
Article 21. Limitation of liability,

DOI: 10.4018/978-1-7998-7407-2.ch013

Article 23. Offences and penalties,
Article 24. Entering into and ratification of conventions, treaties and agreements.
Article 26. State bound, This Act shall bind the State, except in so far as criminal liability is concerned.
Article 27. Short title and commencement, This Act shall be called the Space Affairs Act, 1993, and shall come into operation on a date fixed by the State President by proclamation in the Gazette.

13.1.2. The South African National Space Agency

The South African National Space Agency (SANSA) is South Africa's government agency responsible for the promotion and development of aeronautics and aerospace space research. It fosters cooperation in space-related activities and research in space science, seeks to advance scientific engineering through human capital, as well as the peaceful use of outer space, and supports the creation of an environment conducive to the industrial development of space technologies within the framework of national government. SANSA was established on 9 December 2010 by the National Space Agency Act. Currently, SANSA's main focusses include using data obtained from remote sensing through satellites and other projects to provide assessment on flooding, fires, resource management and environmental phenomena in South Africa and the African continent.[2]

SANSA was formed after an act of parliament was passed by acting President Kgalema Motlanthe in 2009.

The agency was formed with the intent of consolidating space-related research, projects and research in South Africa and to assume the role as a regional center for space research in Africa.

Throughout the 1950s to 1970s lunar and interplanetary missions conducted by NASA had been supported from a tracking station at Hartebeesthoek where the first images of Mars were received from the Mariner IV spacecraft in the first successful flyby of the planet.

Other South African facilities also assisted in tracking satellites to determine the effects of the upper atmosphere on their orbits.[3]

In 1980s work on the development of a launcher and a satellite had been in progress but was discontinued after 1994.

In 1999, South Africa launched its first satellite, SUNSAT from Vandenberg Air Force Base in the US. A second satellite, SumbandilaSat, was launched from the Baikonur Cosmodrome in Kazakhstan in 2009.[4]

ADDITIONAL READING

Kim, D. H. (2005). Example of Legislation on the Space Relations of Every Countries in the World and Main Contents of the New Space Exploitation Promotion Act in Korea. *Proceeding of the 35th International Conference on the 21st Century New International Air and Space Legal Oder published by the Korean Association of Air and Space Law*, 68-81.

KEY TERMS AND DEFINITIONS

Jupiter: Jupiter is the fifth planet from the Sun and the largest in the Solar System. It is a gas giant with a mass one-thousandth that of the Sun, but two-and-a-half times that of all the other planets in the Solar System combined. Jupiter is one of the brightest objects visible to the naked eye in the night sky, and has been known to ancient civilizations since before recorded history. It is named after the Roman god Jupiter. When viewed from Earth, Jupiter can be bright enough for its reflected light to cast visible shadows, and is on average the third-brightest natural object in the night sky after the Moon and Venus. Jupiter is primarily composed of hydrogen with a quarter of its mass being helium, though helium comprises only about a tenth of the number of molecules.

Saturn: Saturn is the sixth planet from the Sun and the second-largest in the Solar System, after Jupiter. It is a gas giant with an average radius of about nine times that of Earth. It only has one-eighth the average density of Earth; however, with its larger volume, Saturn is over 95 times more massive. Saturn is named after the Roman god of wealth and agriculture; its astronomical symbol (♄) represents the god's sickle. The Romans named the seventh day of the week Saturday, *Sāturni diēs* ("Saturn's Day") no later than the 2nd century for the planet Saturn.

Venus: Venus is the second planet from the Sun. It is named after the Roman goddess of love and beauty. As the second-brightest natural object in Earth's night sky after the Moon, Venus can cast shadows and can be, on rare occasion, visible to the naked eye in broad daylight. Venus lies within Earth's orbit, and so never appears to venture far from the Sun, either setting in the west just after dusk or rising in the east a bit before dawn. Venus orbits the Sun every 224.7 Earth days. With a rotation period of 243 Earth days, it takes longer to rotate about its axis than any other planet in the Solar System by far, and does so in the opposite direction to all but Uranus (meaning the Sun rises in the west and sets in the east).

ENDNOTES

1 https://www.unoosa.org/oosa/en/ourwork/spacelaw/nationalspacelaw/south_africa/space_affairs_act_1993E.html

2 Campbell, Keith. *"South Africa's space agency advances new satellite and use of imagery"*. Engineering News. Retrieved 15 November 2017.

3 *"Why a South African national space agency?"* (PDF). Engineering News Online. 6 February 2009. Retrieved 19 June 2011.

4 "SA satellite finally lifts off". News24. 18 August 2009. Archived from the original on 22 September 2009.

Chapter 14
Activities of the International Space Organization and Station

ABSTRACT

This chapter describes the establishment process, purpose of establishment, mission, exploration plan, activities of the European Space Agency (ESA) and International Space Station (ISS), and an explanation of the contents of the treaty that is legal basis for its establishment. The European Space Agency (ESA) is an intergovernmental organization of 22 member states dedicated to the exploration of space. Established in 1975 and headquartered in Paris, France, ESA has a worldwide staff of about 2,200 in 2018 and an annual budget of about € 6.68 billion (US $ 7.43 billion) in 2020. ESA also works closely with space organizations outside Europe. ESA has missions planned for Jupiter (JUICE, 2022) and others that will seek dark matter (Euclid, 2020) and observe the energetic universe (Athena, 2028). The International Space Station (ISS) is a space station (habitable artificial satellite) in low Earth orbit. The ISS programme is a joint project between five participating space agencies: NASA (United States), Roscomos (Russia), JAXA (Japan), ESA (Europe), and CSA (Canada).

14.1. EUROPEAN SPACE AGENCY

14.1.1. Outline of European Space Agency

The European Space Agency (ESA) is an intergovernmental organization of 22 member states dedicated to the exploration of space. Established in 1975 and headquartered in Paris, France. ESA has a worldwide staff of about 2,200 in 2018 and an annual budget of about € 6.68 billion (US $ 7.43 billion) in 2020.[2] ESA's space flight programme includes human space flight (mainly through participation in the International Space Station program); the launch and operation of unscrewed exploration missions to other planets and the Moon; Earth observation, science and telecommunication; designing launch vehicles; and maintaining a major space [port the Guiana Space Centre at Kourou, French Guiana.

DOI: 10.4018/978-1-7998-7407-2.ch014

Figure 1.
Source: https://www.esa.int/About_Us/Corporate_news/Member_States_Cooperating_States

ESA Member States: Austria, Belgium, Czech Republic, Denmark, Estonia, Finland, France, Germany, Greece, Hungary, Ireland, Italy, Luxembourg, The Netherlands, Norway, Poland, Portugal, Romania, Spain, Sweden, Switzerland and the United Kingdom.

ESA member states is the European 22 countries. Non-full members, Canada and Slovenia.

ESA also works closely with space organizations outside Europe. ESA currently has missions planned for Mercury (BepiColombo, 2018) and Jupiter (JUICE, 2022), and others that will seek dark matter (Euclid, 2020) and observe the energetic universe (Athena, 2028).[3]

The European Space Agency's facilities are distributed among the following centers: [4]

- ESA science missions are base ESTEC in Noordwijk, Netherlands;
- Earth Observation missions at ESA Centre for Earth Observation in Frascati, Italy;
- ESA Mission Control (ESOC) is in Darmstadt, Germany;
- the European Astronaut Centre (EAC) that trains astronauts for future
- missions is situated in Cologne, Germany; the European Centre for Space Applications and Telecommunications
- (ECSAT), a research institute created in 2009, is located in Harwell, England; and the European Space Astronomy Centre (ESAC) is located in
- Villanueva de la Canada, Madrid, The European Space Agency Science Programme is a long-term programme of space science and space exploration missions.

14.1.2. ESA's Lunar Mission

More than 50 years after the first human landed on the Moon, what is next? The exploration of our Moon is a global endeavor with Europeans and commercial actors playing a large role. The European Space Agency has developed an exploration programme based on four main missions.

Luna Resurs is a partnership with the Russian agency Roscosmos that will carry European technology to land precisely and safely on the Moon (PILOT) and to extract and analyze samples of the lunar terrain (PROSPECT). Orion and the European Service Module will return humans to the Moon and take advantage of the new technology for human space transportation.

Orion, the NASA spacecraft, will bring humans farther than they have never been before. ESA is providing the service modules that will provide propulsion, life support, power, air and water, and control the temperature in the crew module. "In-situ resource utilization (ISRU)"[5] aims to extract and process resources on the Moon into useful products and services. A mission to explore lunar resources could be a reality from 2025. The goal is to produce drinkable water or breathable oxygen on the Moon.

The Heracles mission could take of in 2028 to allow us to gain knowledge on human-robotic interaction while landing a spacecraft on the Moon, collecting samples with a rover operated from the lunar Gateway and send samples back to Earth.[6]

14.1.3. ESA's Exploration on the Mars Atomsheres

While life arose and evolved on Earth, Mars experienced serious climate change.

Planetary geologists can study rocks, sediments and soils for clues to uncover the history of the surface. Scientists are interested in the history of water on Mars to understand how life could have survived. Volcanoes, craters from meteoroid impacts, signs of atmospheric or photochemical effects and geophysical processes all carry aspects of Mars' history. Samples of the atmosphere could reveal crucial details on its formation and evolution, and also why Mars has less atmosphere than Earth.

Figure 2.
Sources: https://www.esa.int/Science_Exploration/Human_and_Robotic

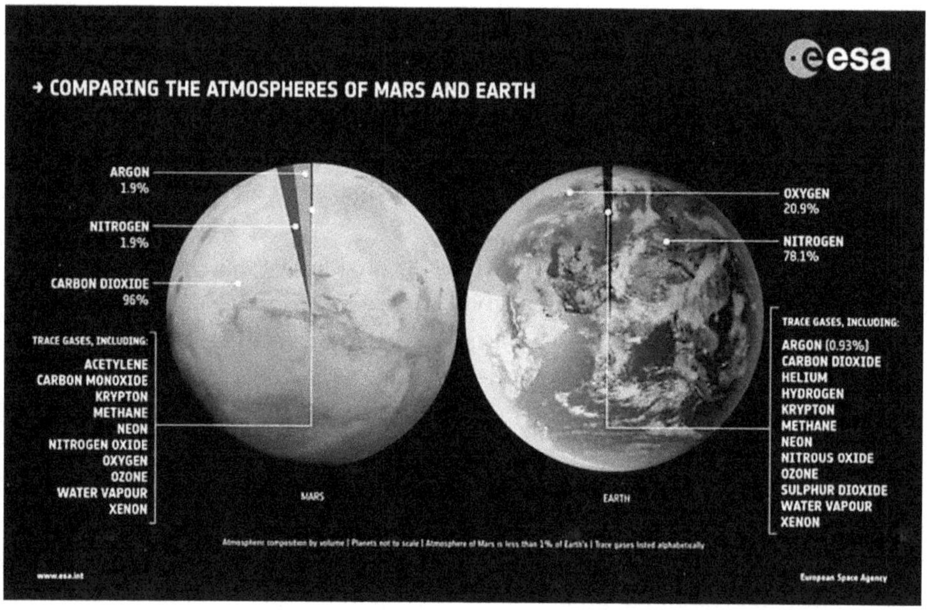

Mars can also help us to learn more about our home. Understanding Martian geophysical processes promises to uncover details of the evolution and history of Earth and other planets in our Solar System.

14.1.4. ESA's Exploration on the Mars Sample-Return

A Mars Sample-Return (MSR) mission would be a spaceflight mission to collect rock and dust samples on Mars and then return them to Earth. Sample-return would be a very powerful type of exploration, because the analysis is freed from the time, budget, and space constraints of spacecraft sensors.

Figure 3.
Source: https://www.planetary.org/blogs/guest-blogs/2019/nasa-esa-latest-msr-plan.html

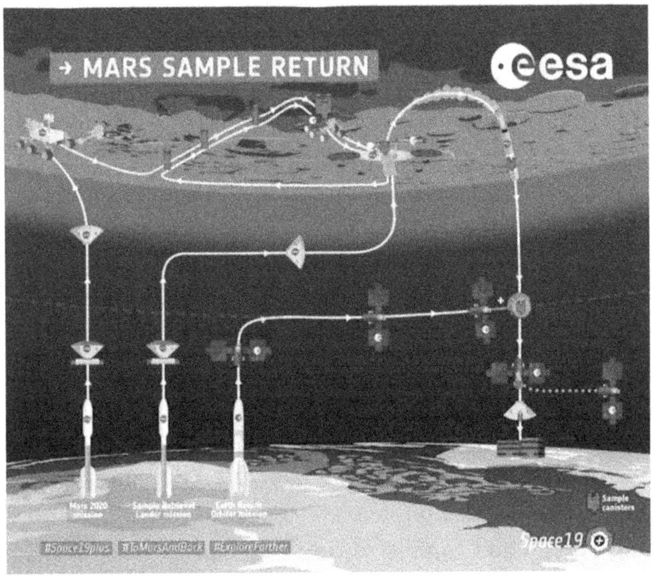

According to Louis Friedman, Executive Director of The Planetary Society, a Mars sample-return mission is often described by the planetary science community as one of the most important robotic space missions, due to its high expected scientific return on investment and its ability to prove the technology needed for a human mission to Mars.

Over time, several concept missions have been studied, but none of them got beyond the study phase. The three latest concepts for an A Mars Sample-Return (MSR) mission are a NASA-ESA proposal, a Russian proposal (Mars-Grunt), and a Chinese proposal.

The current strategy for MSR is an international mission with primary mission hardware provided by NASA and ESA. Additional international support for the mission is being solicited, but these collaborations are still in early negotiations.

According to JPL's Chad Edwards, the goal was to "break up the mission into affordable, technically achievable, cleanly-defined segments," adding "nothing needs a technological miracle to make this happen.

Figure 4.
Source: https://www.nasa.gov/sites/default/files/thumbnails/image/pia12114_0.jpg

"The mission segments consist of the Mars 2020 rover, a solar-powered surface retrieval lander and rover mission jointly developed between NASA and ESA, and a sample return orbiter built by ESA with NASA contributions.

NASA provided input on ESA's Earth return orbiter and sample retrieval rover design studies, while ESA provided input on updates to the general sample retrieval lander design concept. Watzin noted the crucial importance of this international collaboration, saying "it's important that if this mission is to succeed that NASA and ESA work as one big team."

the Earth-return orbiter (ERO) and sample retrieval lander/fetch rover (SRL) will depart Earth in 2026, in different months. MSR is an international mission with primary mission hardware provided by NASA and ESA. Additional international support for the mission is being solicited, but these collaborations are still in early negotiations.

According to JPL's Chad Edwards, the goal was to "break up the mission into affordable, technically achievable, cleanly-defined segments," adding "nothing needs a technological miracle to make this happen". This overview of the ESA–NASA Mars Sample Return mission shows launches of the Mars 2020 rover, the Sample Retrieval Lander, and Earth Return Orbiter.

The Mars 2020 rover collects samples and leaves them in canisters on the surface, and possibly visits the retrieval lander directly. The lander deploys a fetch rover to collect the Mars 2020 samples and deposit them in an ascent vehicle, which blasts into Mars orbit.

There, a return orbiter collects the samples for transport back to Earth. The mission segments consist of the Mars 2020 rover, a solar-powered surface retrieval lander and rover mission jointly developed between NASA and ESA, and a sample return orbiter built by ESA with NASA contributions.

NASA provided input on ESA's Earth return orbiter and sample retrieval rover design studies, while ESA provided input on updates to the general sample retrieval lander design concept. Watkin noted the crucial importance of this international collaboration, saying "it's important that if this mission is to succeed that NASA and ESA work as one big team. "Both the Earth-return orbiter (ERO) and sample retrieval lander/fetch rover (SRL) will depart Earth in 2026, in different months.

14.1.5. ESA's Jupiter Icy Moons Explorer Mission[7]

The Jupiter Icy moons Explorer (JUICE) is an interplanetary spacecraft in development by the ESA with Airbus Defense and Space as the main contractor. The mission will study three of Jupiter's Galilean moons: Ganymede, Callisto, and Europa (excluding the more volcanically active Io) all of which are thought to have significant bodies of liquid water beneath their surfaces, making them potentially habitable environments.[8]

The spacecraft is set for launch in June 2022 and will reach Jupiter in October 2029 after five gravity assists and 88 months of travel.

By September 2032 the spacecraft will enter orbit around Ganymede for its close up science mission and becoming the first spacecraft to orbit a moon other than the moon of Earth. The selection of this mission for the L1 launch slot of ESA's Cosmic Vision science programme was announced on 2 May 2012.

Its period of operations will overlap with NASA's *Europa Clipper* mission, launching in 2024. JUICE will be launched in June 2022 by an Ariane 5 rocket. Following the launch, there is a planned first flyby of Earth in May 2023, Venus in October 2023, second flyby of Earth in September 2024, Mars in February 2025 and a final third flyby of Earth in November 2026, to put the JUICE on a trajectory to Jupiter. In October 2029, when it arrives in Jupiter's system, JUICE will first perform a flyby of Ganymede in preparation for orbital insertion ≈7.5 hours later.

The first orbit will be elongated, with the first Jupiter closest approach taking place in May 2030. After that, the orbits will be gradually closer to Jupiter, resulting in a circular orbit.

The first Europa flyby will take place in October 2030. JUICE will enter a high inclination orbit to allow exploration of Jupiter's polar regions. JUICE will study Jupiter's magnetosphere. Then a Calisto flyby in April 2031 will put JUICE on a normal equatorial orbit. Also, there is a transit of Europa and Io that will occur on 27 January 2032.[9]

14.1.6. ESA's ARIEL Space Mission

The ARIEL (Atmospheric Remote-sensing Exoplanet Large-survey) mission was selected by the European Space Agency (ESA) for its next medium-class science mission due for launch in 2028. The goal of the ARIEL mission is to investigate the atmospheres of several hundreds planets orbiting distant stars in order to address the fundamental questions on how planetary systems form and evolve.

During its 4-year mission, ARIEL will observe 1000 planets exoplanets ranging from Jupiter- and Neptune-size down to super-Earth size in the visible and the infrared with its meter-class telescope.

The analysis of ARIEL spectra and photometric data will allow to extract the chemical fingerprints of gases and condensates in the planets' atmospheres, including the elemental composition for the most favorable targets.

It will also enable the study of thermal and scattering properties of the atmosphere as the planet orbit around the star.[10]

14.1.7. ESA's Moon Village Construction

The European Space Agency (ESA) plans to create a space station on the surface of the moon. ESA's historic Rosetta mission[11] concluded as planned, on 30 September 2016, with a controlled impact onto the comet it had been investigating for more than two years. The mission was launched on 2 March 2004, on a 10-year journey towards comet 67P/Churyumov-Gerasimenko.

En route, it passed by two asteroids, 2867 Steins (in 2008) and 21 Lutetia (in 2010), before entering deep-space hibernation mode in June 2011. On 20 January 2014, it 'woke up' and prepared for arrival at the comet in August that year. On 12 November, the mission deployed its Philae probe to the comet, the first time in history that such an extraordinary feat was achieved.

During the next phase of the mission, Rosetta accompanied the comet through perihelion (13 August 2015) until the end of the mission. On November 12, 2014, the lander "Philae" mounted on the European Space Agency (ESA) comet explorer "Rosetta" landed on Comet Churymov-Gerasimenko.

This is the first time a spacecraft has landed on a comet. ESA's Philae was the first spacecraft to land on a comet at the historic moment of space exploration.

The European Space Agency is planning for a lunar resource exploration by 2025. ESA director Jan Wörner hopes that the "moon village" will function properly for 20 years.

Figure 5. European Space Agency's Moon Village
Source: https://blogs.esa.int/janwoerner/2016/11/23/moon-village

The ESA further stated that building a permanent base on the moon is a "stepladder" and "testbed" for a visit to Mars. ESA is planning to install a permanent human outpost on the moon. This imagined "moon village" is the product of international cooperation between countries far from the earth, and how to build a lunar colony in the future, space science, space business, and space mining.

And it will be the base for space tourism. Although a lunar sediment called "regolith" covers the surface of the moon and has been deposited at least 12 feet (3 m 65 cm) deep underground, ESA has a plan to mine this lunar sediment in the future.

These sedimentary layers are a mixture of compounds such as clay, glass fragments, minerals, and iron oxide, and it is considered that oxygen, water, and fuel can be extracted from iron oxide.

Moon Village will be made available by ESA member states and other countries around the world, Dr. Johann-Dietrich Woerner (Director General of ESA) said.

He added that the ESA should make the moon the next destination for humans in low-Earth orbit, use the moon close to Earth, and pave the way for Mars.[12]

Moon Village Association (MVA) was established in 2017, as a non-governmental organization based in Vienna, whose goal is the creation of a permanent global informal forum for stakeholders like governments, industry, academia and the public interested in the development of the Moon Village.

The MVA fosters cooperation for existing or planned global moon exploration programs, be they public or private initiatives. It comprises approximately 220 members from more than 39 countries and 26 Institutional members around the globe, representing a diverse array of technical, scientific, cultural and interdisciplinary fields.

14.2. INTERNATIONAL SPACE STATION

14.2.1. Outline of the International Space Station

The International Space Station (ISS) is a space station (habitable artificial satellite) in low Earth orbit. The ISS programme is a joint project between five participating space agencies: NASA (United States), Roscomos (Russia), JAXA (Japan), ESA (Europe), and CSA (Canada). The ownership and use of the space station is established by intergovernmental treaties and agreements. [13]

The ISS serves as a microgravity and space environment research laboratory in which scientific experiments are conducted in astrobiology, astronomy, meteorology, physics, and other fields. The station is suited for testing the spacecraft systems and equipment required for possible future long-duration missions to the Moon and Mars.[14]

It is the largest artificial object in space and the largest satellite in low Earth orbit, regularly visible to the naked eye from Earth's surface.

It maintains an orbit with an average altitude of 400 kilometers (250 mi) by means of reboots man oeuvres using the engines of the *Zvezda* Service Module or visiting spacecraft. The ISS circles the Earth in roughly 92 minutes, completing 15.5 orbits per day.[15] The station is divided into two sections: the Russian Orbital Segment (ROS), operated by Russia; and the United States Orbital Segment (USOS), which is shared by many nations.

Roscosmos has endorsed the continued operation of ISS through 2024,[16] but had previously proposed using elements of the Russian segment to construct a new Russian space station called OPSEK.[17] As of December 2018, the station is expected to operate until 2030. [18]The first ISS component was launched in 1998, with the first long-term residents arriving on 2 November 2000. Since then, the station has been continuously occupied for 19 years and 188 days.

As of September 2019, 239 people from 19 countries had visited the space station, many of them multiple times. The United States sent 151 people, Russia sent 47, nine were Japanese, eight Canadian, five Italian, four French, three German, and one each from Belgium, Brazil, Denmark, Kazakhstan, Malaysia, the Netherlands, South Africa, the Republic of Korea, Spain, Sweden, the United Arab Emirates, and the United Kingdom.[19]

As of November 2020, 242 astronauts, cosmonauts, and space tourist from 19 different nations have visited the space station, many of them multiple times; this includes 152 Americans, 49 Russians, 9 Japanese, 8 Canadians, and 5 Italians.[20]

On 28 March 2015, Russian sources announced that Roscosmos and NASA had agreed to collaborate on the development of a replacement for the current ISS.[21]

Mr. Igor Komarov, the head of Russia's Roscosmos, made the announcement with NASA administrator Charles Bolden at his side.

In a statement provided to Space News on 28 March, NASA spokesman David Weaver said the agency appreciated the Russian commitment to extending the ISS, but did not confirm any plans for a future space station.[22]

Figure 6. International Space Station (ISS) Participating Countries

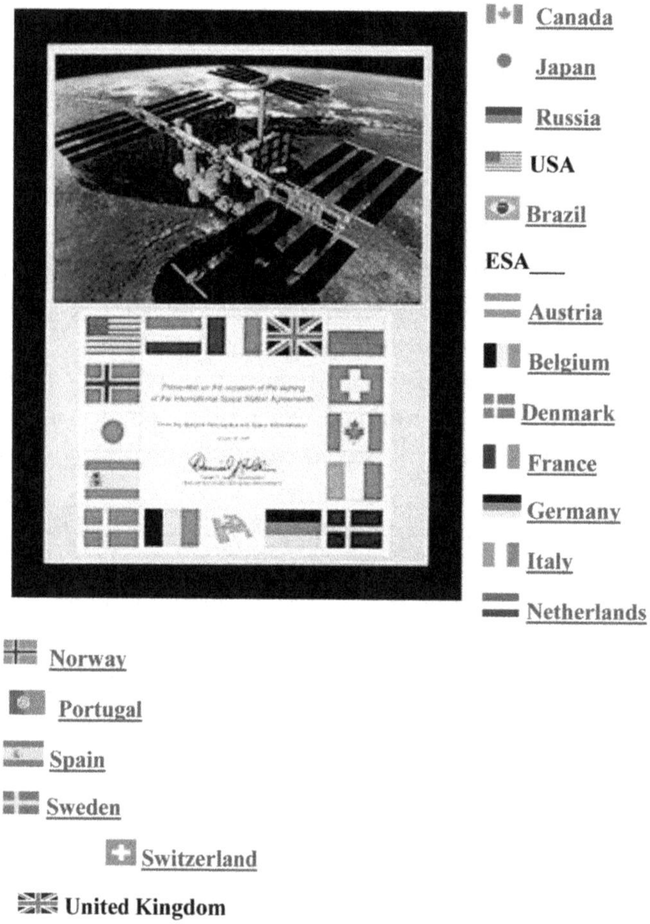

The ISS consists of Canada, Japan, the Russian Federation, The United States, and eleven Member States of the European Space Agency (Belgium,Denmark, France, Germany, Italy, The Netherlands, Norway, Spain, Sweden, Switzerland and The United Kingdom.

14.2.2. The International Space Station Legal Framework

The International Space Station is a co-operative programme between Europe, the United States, Russia, Canada, and Japan for the joint development, operation and utilization of a permanently inhabited Space Station in low Earth orbit. The legal framework defines the rights and obligations of each of the countries and their jurisdiction and control with respect to their Space Station elements.

- **The Space Station Agreements:** The International Space Station legal framework is built on three levels of international co-operation agreements.
- The International Space Station Intergovernmental Agreement, often referred to as 'the IGA', is an international treaty signed on 29 January 1998 by the fifteen governments involved in the Space Station project. This key government-level document establishes 'a long term international co-operative frame-work on the basis of genuine partnership, for the detailed design, development, operation, and utilization of a permanently inhabited civil Space Station for peaceful purposes, in accordance with international law' (Article 1);
- Four Memorandum of Understandings (MoUs) between the National Aeronautics and Space Administration (NASA) and each co-operating Space Agency: European Space Agency (ESA), Canadian Space Agency (CSA), Russian Federal Space Agency (Roscosmos), and Japan Aerospace Exploration Agency (JAXA);.
- The objective of these space agencies-level agreements is to describe in details the roles and responsibilities of the agencies in the design, development operation and utilization of the Station. In addition, the agreements serve to establish the management structure and interfaces necessary to ensure effectively the utilization of the Station;
- Various bilateral Implementing Arrangements between the space agencies have been established to implement the Memoranda of Understandings. The Arrangements distribute concrete guidelines and tasks among the national agencies.
- The Intergovernmental Agreement, establishing the International Space Station cooperative framework, has been signed by fourteen governments: the United States of America, Canada, Japan, the Russian Federation, and 10 Member States of the European Space Agency (Belgium, Denmark, France, Germany, Italy, The Netherlands, Norway, Spain, Sweden and Switzerland; the UK joined in 2012, with Hungary and Luxembourg committing to the Space Station programme in 2019 as well as ESA cooperating state Slovenia).

14.2.3. Extension of the International Space Station

On 30 September 2015, Boeing's contract with NASA as prime contractor for the ISS was extended to 30 September 2020. Part of Boeing's services under the contract will relate to extending the station's primary structural hardware past 2020 to the end of 2028.[23]

Regarding extending the ISS, on 15 November 2016 General Director Vladimir Solntsev of RSC Energia stated "Maybe the ISS will receive continued resources

Today we discussed the possibility of using the station until 2028," with discussion to continue under the new presidential administration.[24]

As of December 2018, the station is funded only until 2025 and may be de-orbited in 2030. There have also been suggestions that the station could be converted to commercial operations after it is retired by government entities.[25]

In July 2018, the Space Frontier Act of 2018 was intended to extend operations of the ISS to 2030.

This bill was unanimously approved in the Senate, but failed to pass in the U.S. House.[26] In September 2018, the Leading Human Spaceflight Act was introduced with the intent to extend operations of the ISS to 2030, and was confirmed in a law prohibiting contact between US and Chinese space programs. As of 2019, China is not involved in the International Space Station.

In addition to national security concerns, United States objections include China's human rights record and issues surrounding transfer.[27] The heads of both the South Korean and Indian space agencies announced at the first plenary session of the 2009 International Astronautical Congress on 12 October that their nations intend to join the ISS program.

14.2.4. NASA has Subcontracted a Private Company to Manufacture Three Modules for the International Space Station

NASA has selected a private company to deliver up to three new modules to the International Space Station (ISS), which may form their own commercial space station when the ISS is retired. Axiom Space from Houston, Texas has been awarded a contract by the US space agency to launch its first module in late 2024, a central "node" module, with two more to follow.

While beginning life attached to the station, the company says these modules could later detach and form a "replacement" for the ISS.

The two additional modules will comprise a habitat for the crew and a research and manufacturing module, which will contain a large window similar to the station's current Cupola module. This will give the astronauts on board extra room to live and work in, and potentially provide new destinations for commercial astronauts.

ADDITIONAL READING

Cruz, T. (2018). *S.3277 - 115th Congress (2017-2018): Space Frontier Act of 2018*. www.congress.gov

Foust, J. (2018). *House joins Senate in push to extend ISS*. Space News.

Oberg. (2001). *China takes aim at the space station*. NBC News.

KEY TERMS AND DEFINITIONS

Atomsheres of Mars: The Atmosphere of Mars is the layer of gases surrounding Mars. It is primarily composed of carbon dioxide (95.32%), molecular nitrogen (2.6%) and argon (1.9%). It also contains trace levels of water vapor, oxygen, carbon monoxide, hydrogen and other noble gases.The atmosphere of Mars is much thinner than Earth's. The surface pressure is only about 610 pascals (0.088 psi) which is less than 1% of the Earth's value. The currently thin Martian atmosphere prohibits the existence of liquid water at the surface of Mars, but many studies suggest that the Martian atmosphere was much thicker in the past. The atmosphere of Mars has been losing mass to space throughout history, and the leakage of gases still continues today. The atmosphere of Mars is colder than Earth's.

ESA's ARIEL Space Mission: Ariel, the Atmospheric Remote-sensing Infrared Exoplanet Large-survey, was selected as the fourth medium-class mission in ESA's Cosmic Vision programme. During its 4-year mission, Ariel will study what exoplanets are made of, how they formed and how they evolve, by surveying a diverse sample of about 1000 extrasolar planets, simultaneously in visible and infrared wavelengths. It is the first mission dedicated to measuring the chemical composition and thermal structures of hundreds of transiting exoplanets, enabling planetary science far beyond the boundaries of the Solar System.

Jupiter Icy Moons Explorer (JUICE): The Jupiter Icy moons Explorer (JUICE) is an interplanetary spacecraft in development by the European Space Agency (ESA) with Airbus Defence and Space as the main contractor. The mission will study three of Jupiter's Galilean moons: Ganymede, Callisto, and Europa (excluding the more volcanically active Io) all of which are thought to have significant bodies of liquid water beneath their surfaces, making them potentially habitable environments. The spacecraft is set for launch in June 2022 and will reach Jupiter in October 2029 after five gravity assists and 88 months of travel. By September 2032, the spacecraft will enter orbit around Ganymede for its close up science mission, becoming the first spacecraft to orbit a moon other than the moon of Earth. The selection of this mission for the L1 launch slot of ESA's Cosmic Vision science programme was announced on 2 May 2012. Its period of operations will overlap with NASA's *Europa Clipper* mission, launching in 2024.

Luna Resurs: According to the original plan, Luna-Resurs was to be the second post-Soviet robotic mission to the Moon. At one point, Luna-Resurs was a joint project with India, but, eventually the European Space Agency, ESA took the role of the main international partner in the project, which had a main goal of examining the distribution of volatilities, such as water, in the lunar surface, which would be crucial for both, the scientific understanding of the Moon and for the future of space exploration.

Mars Sample-Return (MSR): Mars Sample Return is a proposed mission to return samples from the surface of Mars to Earth. The mission would use robotic systems and a Mars ascent rocket to collect and send samples of Martian rocks, soils, and atmosphere to Earth for detailed chemical and physical analysis.

ENDNOTES

1 *"ESA - Funding"*. esa.int. 15 February 2020.
2 *"ESA - Funding"*. esa.int. 15 February 2020.
3 https://www.space.com/22562-european-space-agency.html
4 https://en.wikipedia.org/wiki/European_Space_Agency
5 https://www.sciencedirect.com/topics/earth-and-planetary-sciences/in-situ-resource-utilization
6 https://www.esa.int/Education/Teach_with_the_Moon/Lunar_Exploration_ESA_s_missions
7 https://en.wikipedia.org/wiki/Jupiter_Icy_Moons_Explorer
8 *"ESA—Selection of the L1 mission"* (PDF). 17 April 2012.
9 https://en.wikipedia.org/wiki/Jupiter_Icy_Moons_Explorer
10 https://www.ucl.ac.uk/exoplanets/missions/ariel-space-mission
11 https://sci.esa.int/web/rosetta/

12 https://www.theguardian.com/science/2016/sep/23/is-a-moon-village-the-next-step-for-space-exploration-esas-chief-thinks-so

13 "*International Space Station legal framework*". European Space Agency (ESA). 19 November 2013. Retrieved 21 February 2015.

14 "*ISS Research Program"*. NASA. Archived from the original on 13 February 2009. Retrieved 27 February 2009.

15 *"Current ISS Tracking data"*. NASA. 15 December 2008. Retrieved 28 January 2009. This article incorporates text from this source, which is in the public domain.

16 de Selding, Peter B. (25 February 2015). *"Russia — and Its Modules — To Part Ways with ISS in 2024"*. Space News. Retrieved 26 February 2015.

17 Bodner, Matthew (17 November 2014). *"Russia May Be Planning National Space Station to Replace ISS"*. The Moscow Times. Retrieved 3 March 2015.

18 Nelson, Senator Bill (20 December 2018). "The Senate just passed my bill to help commercial space companies launch more than one rocket a day from Florida! This is an exciting bill that will help create jobs and keep rockets roaring from the Cape. It also extends the International Space Station to 2030!".

19 Visitors to the Station by Country NASA, 25 September 2019.

20 Visitors to the Station by Country NASA, 25 September 2019. *This article incorporates text from this source, which is in the public domain.*

21 "*Russia & US agree to build new space station after ISS, work on joint Mars project*". RT.com. March 28, 2015. Retrieved March 28, 2015.

22 Foust, Jeff (March 28, 2015). *"NASA Says No Plans for ISS Replacement with Russia"*. Space News.

23 Maass, Ryan (September 30, 2015). *"NASA extends Boeing contract for International Space Station"*. Space Daily. UPI. Retrieved October 2, 2015.

24 "ISS' Life Span Could Extend Into 2028 – Space Corporation Energia Director". Sputnik. November 15, 2016. Retrieved November 18, 2016.

25 Grush, Loren (January 24, 2018). *"Trump administration wants to end NASA funding for the International Space Station by 2025"*. The Verge. Retrieved April 24, 2018.

26 Cruz, Ted (December 21, 2018). "S.3277 - 115th Congress (2017-2018): Space Frontier Act of 2018". www.congress.gov. Retrieved March 18, 2019.

27 James Oberg (October 26, 2001). *"China takes aim at the space station"*. NBC News. Retrieved January 30, 2009.

Chapter 15
EU Draft Code of Conduct for Outer Space Activities:
Space Debris and Liability Convention

ABSTRACT

This chapter describes the main contents of the European Union (EU) Draft Code of Conduct for Outer Space Activities, Space Debris and Liability Convention. It is necessary and desirable for us to legislate more concretely the abovementioned draft Code of Conduct for Outer Space Activities so as to mitigate or remove space debris. In December 2008, the EU adopted a Draft Code of Conduct (EU Draft CoC) for outer space activities. On September 2010, the EU revised a second draft of the EU Code of Conduct for Outer Space Activities. The draft of the International Code of Conduct (the Code), dated March 31, 2014, was intended to be the subject of negotiations at the United Nations in New York from July 27-31, 2015. The code, a politically and not legally binding document, aims to establish some rules of good conduct for outer space activities. The author proposes the establishment of a new Asian-Pacific International and Environmental Monitoring Organization (tentative title) for prevention and mitigation of space debris.

15.1. CONTEXT

In December 2008 the EU adopted a Draft Code of Conduct (EU Draft CoC) for outer space activities. On September 2010, the EU revised a second draft of the EU Code of Conduct for Outer Space Activities. The Code, a politically and not legally binding document, aims to establish some rules of good conduct for outer space activities.

Now the European Union will have to try to muster enough adoption of the Code, at the international level, to make it an effective soft law tool for space governance.[1]

According to some experts, many look at the Hague Code of Conduct against Ballistic Missiles Proliferation (HCoC) as a successful example of how "soft law" can be implemented and play a concrete role in an international security context.

DOI: 10.4018/978-1-7998-7407-2.ch015

HCoC has reached in less than 6 years an important number of adhesions: 130 countries including the United States and Russia, which are the most active actors in terms of the development and launch of Ballistic Missiles and Space Launch Vehicles. the large number of subscribing states, the HCoC suffers from a paradox in the sense that significant countries are still absent, such as active actors like DPRK[2], Iran, China and emerging actors in the space sector like Brazil, Mexico and Saudi Arabia.

Most of the countries located in the region going from the north of Africa to eastern Asia do not participate in the HCoC.[3]

The main aspects considered here include the universalization of the EU Draft CoC and its chances to motivate a larger participation in it, its effective implementation by adhering states and the scope of the Code.

In many respects, the Code's provisions are hardly controversial, and reflect commonsense practices for operating in orbit. While critiquing various details about the EU Code, particularly when compared with a more restrictive proposed space arms control treaty, known by the acronym PPWT (Prevention of the Placement of Weapons in Outer Space), supported by China and Russia. At present, various activities in Outer Space are governed by a few globally acknowledged treaty mechanisms like the Outer Space Treaty (OST, 1967) and Moon Agreement (1979).

The agreement on the Rescue of Astronauts, the Return of Astronauts and the Return of Objects Launched into Outer Space (Rescue Agreement, 1968) and other agreements such as the Convention on International Liability for Damage Caused by Space Objects (Space Liability Convention, 1972) are two other important treaties in this regard.

The UN has also undertaken various initiatives like the Committee on the Peaceful Uses of Outer Space (COPUOS, 1959) and Prevention of an Arms Race in Outer Space (PAROS, 1982). The UN has produced various General Assembly Resolutions, Meetings, Conference on Disarmament (CD) working papers, etc., on the issue.

The Measures on space debris control and mitigation of the *Revised Draft Code of Conduct for Outer Space Activities* adopted by the EU Council on 8-9 December 2008 and on 27 September 2010 as the Measures followings;

Measures on space debris control and mitigation in order to limit the creation of space debris and reduce its impact in outer space, the Subscribing States commit to:

- refrain from the intentional destruction of any on-orbit space object or other activities which may generate long-lived space debris;
- adopt and implement, in accordance with their own internal processes, the appropriate policies and procedures or other effective measures in order to implement the Space Debris Mitigation Guidelines of the United Nations Committee for the Peaceful Uses of Outer Space as endorsed by UNGA Resolution 62/217.[4]

It is necessary and desirable for us to legislate more concretely the aforementioned *Draft Code of Conduct for Outer Space Activities* so as to mitigate or remove space debris.

15.2. THE DRAFT OF THE INTERNATIONAL CODE OF CONDUCT OF 2014

The *Draft of the International Code of Conduct* (the Code), dated March 31, 2014, was intended to be the subject of negotiations at the United Nations in New York from July 27 to 31, 2015. The meeting consisted of representatives from over 100 countries by invitation of the European Union, but the intent to negotiate the latest draft of the Code was cut short by two procedural moves.[5]

The first procedural concern raised at the meeting pointed out negotiations of the Code had no UN mandate and could not be considered a UN event even though the meeting was held at UN Headquarters. Consequently, the EU had no legal standing to negotiate the text of the Code in the auspices of the UN, which required the Chair to reclassify the purpose of the meeting from a negotiation to a consultation. A second procedural impediment was raised by several delegations that a possible UN General Assembly mandate, which would allow negotiations for the Code to be held under the auspices of the UN, could not be restricted so as to prevent other countries to propose alternative texts of the Code for discussion. These procedural actions occurred against a less than optimistic backdrop for the Code as the prospects of the Code's adoption going into the meeting were slim.

Growing concerns about orbital debris and the risk of collisions has been a driving force behind creating a code of conduct for those operating spacecraft in Earth orbit.

Figure 1.
Source: https://thespacereview.com/article/1794/1

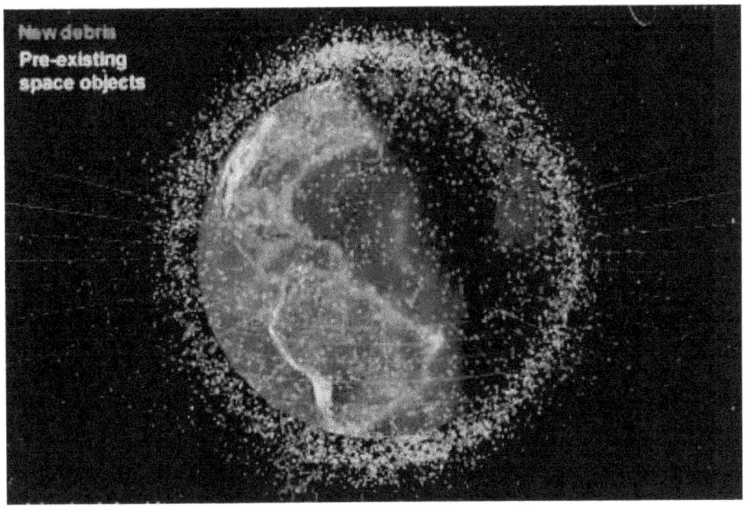

This pall over the Code was exacerbated by the competing interests and input from more than 100 countries invited to the meeting, and was also overshadowed by the continued rejection of the Code by both the Russian Federation and the People's Republic of China (PRC).[6] Aside from the procedural maneuvering, which in effect invalidated the purpose of the meeting, the March 31, 2014, draft of the Code contains significant changes from the September 16, 2013, draft. This essay will examine four of these changes and discuss the Code in general, including concerns moving forward and whether adopting the Code continues to be in the national interests of the United States.[7]

15.3. SPACE DEBRIS AND LIABILITY CONVENTION

Concerns about the harm caused by intentional or accidental events to orbiting space objects have increased in recent years, following two episodes that caused damages to satellites and an impressive amount of debris (Chinese anti-satellite missile test, held on January 2007 and the collision between Iridium 33 satellite (Russia) and Cosmos 2251 satellite (Russia) in February 2009.

The 1972 Liability Convention has not regulated the scope and limits of compensation for damage caused by space debris. The author proposed new framework for innovation on the damage caused by space debris as well as amendment of the Liability Convention.

The space debris problem can only be effectively solved by international cooperation. Consultation and cooperation between space agencies are taking place with increasing frequency. It is my firm opinion that only international and regional cooperation could solve the problem of environmental pollution, including damage caused by space debris.

Meanwhile, States have to keep in mind that the exploration and use outer space is for the benefit and in the interests of all countries. We should try to reach an international agreement, binding on all space-using nations, which encompasses the need to protect the space and earth environment.

The Space Law Committee of ILA was sent it to the UN COPUOS to recommend the legislation of the said new *Draft for the International Instrument.* After the committee of space-faring nations has reached consensus on the basic steps to be taken, and has presented its findings to the Scientific and Technical Subcommittee of the UNCOPUOS, it may be appropriate for the Legal Subcommittee of UNCOPUOS to take up the matter, in order to debate and resolve the definition, jurisdiction and control, liability, and other legal issues that many legal experts consider need addressing in the preparation of an international agreement.

Figure 2. Collision Between Russian Iridium 33 and Kosmos 2251
Source: http://www.spacesafetymagazine.com/space-debris/kessler-syndrome/iridium-33-cosmos-2251-years-later-learned-then

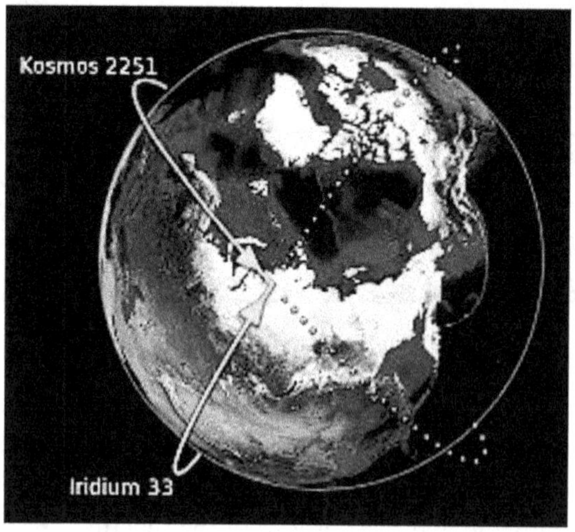

It is necessary for us to exchange the current and various information and view among the countries on the damage caused by space debris and to establish the integrated monitoring system in the world for the purpose of preventing and mitigating the damage caused by space debris as a special and affiliated organization under the United Nations.

First of all, I would like to propose the establishment of a new Asian-Pacific International and Environmental Monitoring Organization (tentative title) for prevention and mitigation of space debris and to track, observe, investigate of space debris and the damage caused by space debris.

Furthermore it is necessary for us to enact independently a the new "*Draft for the International Instrument on the Protection of the Earth Environment from Damage Caused by Space Debris*" as well as to amend some articles of the 1972 Liability Convention in order to protect victims in the case of the damage for compensations caused by the space accidents including space debris.

ADDITIONAL READING

Sachdeva, G. S. (2013). *Outer Space: Law, Policy and Governance*. Amazon.

Rao, Venkata, Gopalakrishnan, V., Abhijeet, Kumar (2017), *Recent Developments in Space Law, Opportunities & Challenges,* 2017, Springer, Unites States

Jakhu, R. S. (2016). *Paul Stephen Dempsey, Routledge Handbook of Space Law*. Routledge Handbooks Online. doi:10.4324/9781315750965

KEY TERMS AND DEFINITIONS

Ballistic Missiles: A ballistic missile follows a ballistic trajectory to deliver one or more warheads on a predetermined target. These weapons are guided only during relatively brief periods—most of the flight is unpowered. Short-range ballistic missiles stay within the Earth's atmosphere, while intercontinental ballistic missiles (ICBMs) are launched on a sub-orbital trajectory. These weapons are in a distinct category from cruise missiles, which are aerodynamically guided in powered flight. An intercontinental ballistic missile trajectory consists of three parts: the powered flight portion; the free-flight portion, which constitutes most of the flight time; and the re-entry phase, where the missile re-enters the Earth's atmosphere. (The flight phases for shorter-range ballistic missiles are essentially the first two phases of the ICBM, as some ballistic categories do not leave the atmosphere.)

Cosmos 2251 Satellite: Kosmos-2251, (Russian: Космос-2251 meaning *Cosmos 2251*), was a Russian Strela-2M military communications satellite. It was launched into Low Earth orbit from Site 132/1 at the Plesetsk Cosmodrome at 04:17 UTC on 16 June 1993, by a Kosmos-3M carrier rocket. The Strela satellites had a lifespan of 5 years, and the Russian government reported that Kosmos-2251 ceased functioning in 1995. Russia was later criticized by *The Space Review* for leaving a defunct satellite in a congested orbit, rather than deorbiting it. In response, Russia noted that they were (and are) not required to do so under international law. In any case, the KAUR-1 satellites had no propulsion system, which may be required for deorbiting. At 16:56 UTC on 10 February 2009, it collided with Iridium 33 (1997-051C), an Iridium satellite, in the first major collision of two satellites in Earth orbit. Iridium satellite, which was operational at the time of the collision, was destroyed, as was Kosmos-2251. NASA reported that a large amount of debris was produced by the collision.

HcoC: The HCoC is an acronym for the Hague Code of Conduct against Ballistic Missile Proliferation (HCoC). It is a widely subscribed multilateral instrument that aims at delegitimizing ballistic missile proliferation. The HCoC consists of a set of general principles, modest commitments, and limited confidence building measures. Hague Code of Conduct against Ballistic Missiles Proliferation. The HCoC aims to contribute to the process of strengthening existing national and international security arrangements and disarmament and non-proliferation objectives and mechanisms. Participants recognize a need to prevent and curb the proliferation of ballistic missile systems capable of delivering weapons of mass destruction, as well as the importance of strengthening, and gaining wider adherence to, multilateral disarmament and non-proliferation mechanisms. The HCoC was formally brought into effect on November 25, 2002, at a launching conference hosted by the Netherlands in The Hague. As of February 2020, 143 countries have subscribed to the HCoC.

Iridium 33 Satellite: Iridium 33 was a communications satellite launched by Russia for Iridium Communications. It was launched into low Earth orbit from Site 81/23 at the Baikonur Cosmodrome at 01:36 UTC on 14 September 1997, by a Proton-K rocket with a Block DM2 upper stage. The launch was arranged by International Launch Services (ILS). It was operated in Plane 3 of the Iridium satellite constellation, with an ascending node of 230.9°. Iridium 33 was part of a commercial communications network consisting of a constellation of 66 LEO spacecraft. The system uses L-Band to provide global communications services through portable handsets. Commercial service begins in 1998.

ENDNOTES

1. Valerio Briani, *The EU as a force for stability in space – The EU Code of Conduct for Outer space Activities,* European Security and Defense Forum Workshop, 11 November 2010; Website; http://www.chathamhouse.org.uk
2. DPRK = Democratic People's Republic of Korea = North Korea.
3. Lucia MARTA, *The Hague Code of Conduct Against Ballistic Missile Proliferation: Lessons Learned" for the European Union Draft Code of Conduct for Outer Space Activities,* ESPI (European Space Policy Institute) No. 34, Perspectives, June, 2010, p.3
4. UN General Assembly, Sixty-second session Agenda item 31, 62/217 International cooperation in the peaceful uses of outer space, 1 February 2008.
5. Tommaso Sgobba, IAASS Statement On International Code of Conduct for Outer Space Operations, Space Safety Magazine, August 5, 2015.
6. Both the Russian Federation and the People's Republic of China continue to promote their draft Treaty on the Prevention of the Placement of Weapons in Outer Space, the Threat or Use of Force against Outer Space Objects (PPWT) in the United Nations Conference of Disarmament. The original draft of the PPWT was presented in 2008 and an updated draft was presented on June 10, 2014, two weeks after the European Union (EU) completed their third Open-Ended Consultations

(OEC) for the development of the International Code of Conduct for Outer Space Activities (ICoC) in Luxembourg at the end of May 2014. The latest draft of the PPWT was rejected by the United States and other nations, but the proposal, along with the Prevention of Arms Race in Outer Space (PAROS), still remain the cornerstone of the Russian Federation and PRC's soft-power strategy.

7 https://www.thespacereview.com/article/2851/1

Chapter 16
Legal Problems on the Compensation for Space Damage Caused by Space Debris and Space Liability Convention

ABSTRACT

This chapter mentions the space debris, observation, space debris' accidents, and the legal problems on the compensation for space damage caused by space debris and relation to the 1972 Space Liability Convention. Space debris is an enemy for mankind. The author comments on the new Draft for the International Instrument on the Protection of the Earth Environment from Damage Caused by Space Debris and presents his proposal on the new legal frame for the prevention and mitigation of space debris and the damage caused by space debris. Frequently, space debris falls back to the Earth, which poses a potential threat to man's exploration and use of outer space activities. Space debris comprise the ever-increasing amount of inactive space hardware in orbit around the Earth as well as fragments of spacecraft that have broken up, exploded, or otherwise become abandoned. It is the author's firm opinion that only international and regional cooperation could solve the problem of environmental pollution including damage caused by space debris.

16.1. INTRODUCTION

The busy space activities of some major space powers, space debris is steadily increasing in quantity and has brought grave potential threats and actual damage to the outer space environment and human activities in space. Frequently, space debris falls back to the earth, which poses a potential threat to man's exploration and use of outer space activities. Space debris comprise the ever increasing amount

DOI: 10.4018/978-1-7998-7407-2.ch016

of inactive space hardware in orbit around the Earth as well as fragments of spacecraft that have broken up, exploded or otherwise become abandoned.

Space debris has become an official enemy of mankind. We should mitigate and remove space debris in the Low Earth Orbit (LEO) and in the Geostationary Orbit (GEO), through international cooperation and agreement in the fields of space science, economics, politics and law in order to safeguard the life and property of mankind and protect the earth's environment. Space debris has created problems, which all space faring nations should endeavor to solve the problems together, in order to protect the environment of earth as well as space for future space development.

The *Convention on International Liability for Damage Caused by Space Objects* (hereinafter referred to Liability Convention) of 1972, also known as the Space Liability Convention, is a treaty that expands on the liability rules created in the Outer Space Treaty of 1967.

The Liability Convention composed of 28 articles. I am going to briefly explain and comment the main contents and key issues on the Liability Convention of 1972 in compare with the legal problems space debris. I would like to give particularly also my view on the legal problems of the *new Draft for the International Instrument on the Protection of the Environment from Damage Caused by Space Debris,*[1]which was proposed by Prof. Dr. Karl Heinz Böckstiegel (Germany) at the Space Law Committee of the aforementioned 66th Buenos Aires Conference of the International Law Association (ILA) on 16 August 1994. Korea is about to join the ranks of advanced countries in the aerospace field.

Korea sent its first woman astronaut into International Space Station (ISS) on April 2008.[2]The Naro (나로) Space Center[3] is also the first space port to be completed in June 11, 2009 at Woinarodo, Goheung country, Junlanam Province on the southern coast of the South Korea. The launch of Korea's first space rocket was "partially successful" as the satellite aboard the rocket failed to reach the intended low earth orbit (LEO) on August 25, 2009.

Korean professors, researchers and experts of space law, members and experts of the Korea Society of Air & Space Law and Policy have gradually become interested in the legal problems relating to the threat of space debris, they have discussed the matter in depth, studying the causes of the accident which create space debris, and the legal problems of the liability for compensation for damage caused by space debris.

16.2. THE OBSERVATION OF SPACE DEBRIS AND ACCIDENTS CAUSED BY SPACE DEBRIS

16.2.1. The Observation of Space Debris

From the mid-1960s until 1990, the launch rate has been fairly constant, at around100-130 per year.[4] Since SPACECOM's detection and tracking capability is limited to 10 cm for low altitude objects and 1m for geosynchronous orbit altitude objects, the number of trackable objects does not include any objects smaller than 10cm 130 per year.[5]

Among these 7,024 trackable objects, functioning satellites constitute only 6% of the total. Since SPACECOM's detection and tracking capability is limited to 10 cm for low altitude objects and 1m for geosynchronous orbit altitude objects, the number of trackable objects does not include any objects smaller than 10cm diameter.[6]

Among these 7,024 trackable objects, functioning satellites constitute only 6% of the total.

The rest is made up of mission-ended satellites (21%), rocket bodies (16%), operational debris (12%) and fragment debris created by break-up in debris (45%). The largest break-up occurred in November 1986, when an Arian upper stage exploded, producing about 500 trackable fragments of which about 60 are still in orbit.[7]

There were 104 break-ups recorded by the end of June 1991. Between the launch of Sputnik on 4 October 1957 and 1 January 2008, approximately 4,600 launches have placed some 6,000 satellites into orbit; about 400 are now travelling beyond the earth on interplanetary trajectories, but of the remaining 5600 only about 800 satellites are operational - roughly 45 percent of these are both in LEO and GEO.

Space debris comprise the ever-increasing amount of inactive space hardware in orbit around the earth as well as fragments of spacecraft that have broken up, exploded or otherwise become abandoned.

About 50 percent of all tractable objects are due to in-orbit explosion events (about 200) or collision events (less than 10). According to US estimates, the amount of debris, including un-tractable objects of more than 1mm in diameter, is 3,500,000 pieces.

Figure 1. Many Dangerous Space Debris Orbiting Around the Satellite
Source: https://www.theverge.com/2019/12/3/20991798/space-debris-low-earth-orbit-satellites-mega-constellations-collisions-graph

The total mass of objects in orbits is 3,000 tons. The orbital velocity of objects in Low Earth Orbit is about 7km/s.[8]

The large number of spacecraft, rocket bodies, and other hardware associated with these missions, encounter one of three fates:

1. re-entry into the earth's atmosphere
2. escape from earth orbit into deep space or
3. remaining in the earth orbit.

After nearly 35 years of international space operations, almost 22,000 space objects, including space debris, have been officially catalogued, with approximately one-third of them still in orbit around the earth. Catalogued' objects are considered to be objects larger than 10-50cmin diameter for LEO and 1 m in diameter in higher orbits, which are sensed and maintained in a database by the United States Space Command's Space Surveillance Network (SSN).[9]

Figure 2. Many Dangerous Space Debris (Trash), Countries and UNOOSA, February, 2021
Source: https://dynamath.scholastic.com/issues/2018-19/040119/space-trash-stats.html#750L

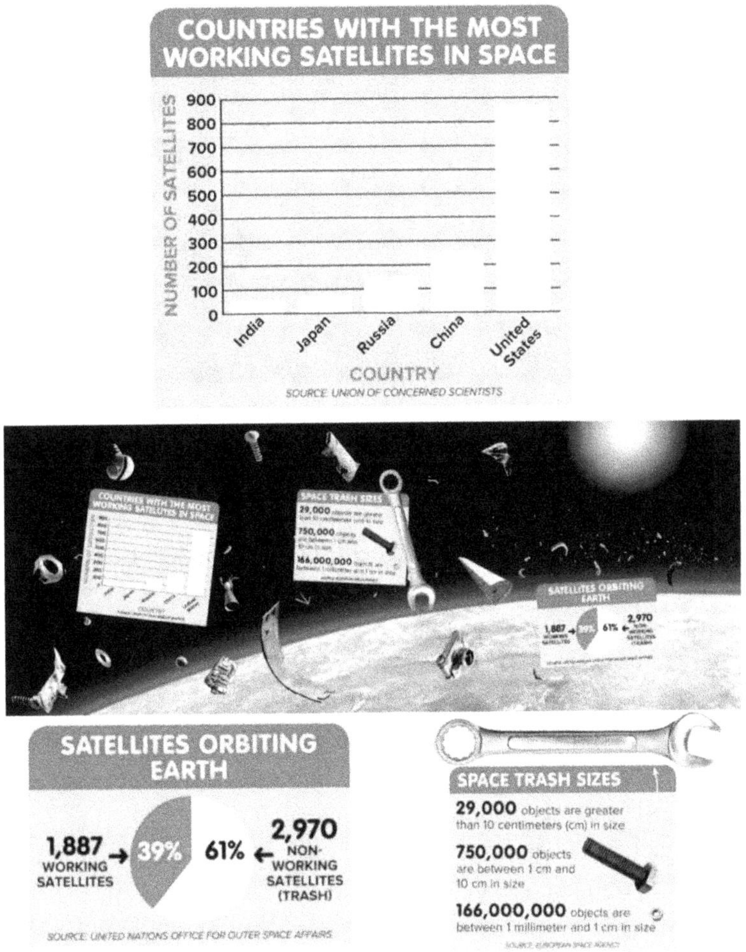

Furthermore, the number of satellites proposed for some of these systems (up to 840 satellites in the case of Teledesic system) raises concerns about their potential contributions to orbital debris in regions of LEO that are already under stress from existing debris.[10]

According to the report of the Japanese Space Debris Study Group, there were about 7,000 pieces of debris of more than 10cm in diameter in space orbit, below an altitude of 5,000km. Recent news announcements about private firms that seek to establish LEO satellite systems for a variety of telecommunications services have generally focused on regulatory and financial issues.

Figure 3. Space Debris are Circling Around the Earth
Source: https://www.nasa.gov/mission_pages/station/news/orbital_debris.html

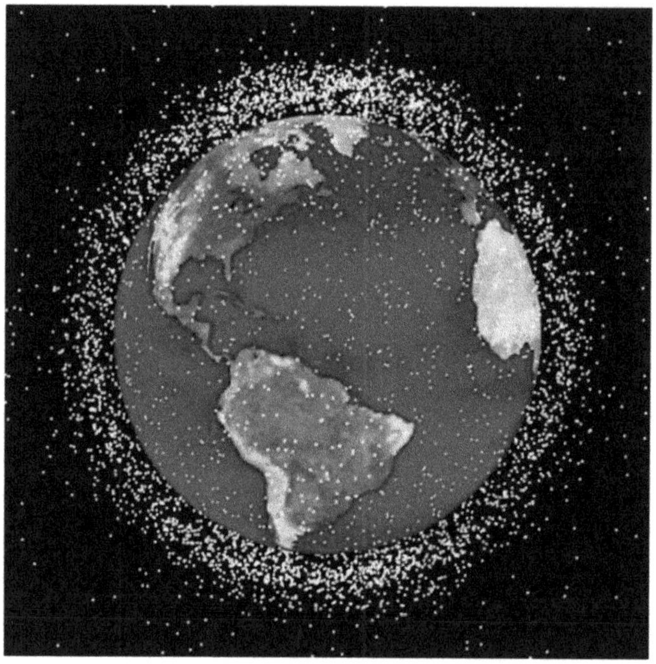

In 1989, ten years later, the North American Aerospace Defense Command (NORAD), which has the capacity to track objects in space from a size of 10cm in diameter at a distance of 500km reported that there had been a total of 19,037 objects launched into outer space, of which some 12,000 had re-entered the earth's atmosphere. In other words, in a period of a decade, there were over 7,671 additional objects launched into outer space. Of those space objects, 95 per cent are now non-functional, that is, uncontrolled, commonly referred to as space junk or debris.

There are thousands of pieces of satellite and other objects which have gone out of control and are in orbit around the earth. According to one source, on average one piece of debris re-enters earth atmosphere every day. While most will burn up on entry into the atmosphere, an increasing number will survive and land on the earth-endangering life and property.

Moreover, as the physical size of space objects increases, so will the probability that a collision will occur between space objects in outer space.

Although there is only a 30 per cent chance of an object hitting land, and a far slimmer chance of that object landing in a populated area, we are not dealing here in the realms of theory, but of reality. The International Space Station operates at about 250 miles altitude, and Space Shuttle flights tend to range between 250 miles and 375 miles. The most debris-crowded area is between 550 miles and 625 miles above the Earth, meaning the risk is less for manned space flight.

The European Space Agency, ESOC, Darmstadt, Germany
Information last updated on 08 January 2021
Number of space debris objects estimated by statistical models to be in orbit
34000 space debris objects greater than 10 cm

900000 space debris objects from greater than 1 cm to 10 cm
128 million space debris objects from greater than 1 mm to 1 cm Source: https://www.esa.int/Safety_Security/Space_Debris/Space_debris_by_the_numbers

On March 27, 2019, India destroyed a weather satellite using an anti-satellite (ASAT) missile, making it the fourth nation after the US, Russia, and China to have conducted such a test.[11] The test was criticized by many for creating space debris (although the Indian Government claimed that this debris would not last in orbit for long).[12]

This action put the growing concern over orbital debris into perspective, since both the criticism and the defense of the ASAT test focused on the creation of space debris and the threat it could pose to the International Space Station.

The European Space Agency estimates that there are in the order of 900,000 pieces of space debris larger than 1 cm in orbit around the earth and over 128 million pieces between 1 cm and 1 mm.

Debris from either size set is capable of causing considerable damage to objects launched from earth and result in the creation of more debris. States Parties also should take measures to avoid harming the environment of the earth through the introduction of extraterrestrial or other matter. Moreover, as the physical size of space objects increases, so will the probability that a collision will occur between space objects in outer space.

16.2.2. Accidents Caused by Space Debris

Although most debris will burn up in the atmosphere, larger objects can reach the ground intact and present a risk. On 12 January 2001, a Star 48 Payload Assist Module AM-D) rocket upper stage re-entered the atmosphere after a "*catastrophic orbital decay*". The PAM-D stage crashed in the sparsely populated Saudi Arabian desert.[13]

It was positively identified as the upper-stage rocket for NAVSTAR 32, a GPS satellite launched in 1993.

The Columbia disaster in 2003 demonstrated this risk, as large portions of the spacecraft reached the ground. In some cases entire equipment systems were left intact.

NASA continues to warn people to avoid contact with the space debris due to the possible presence of hazardous chemicals. On 27 March 2007, wreckage from a Russian spy satellite was spotted by Lan Chile (LAN Airlines) in an Airbus A340, which was travelling between Santiago, Chile, and Auckland, New Zealand carrying 270 passengers. The pilot estimated the space debris was within 8 km of the aircraft, and he reported hearing the sonic boom as it passed.

The aircraft was flying over the Pacific Ocean, which is considered one of the safest places in the world for a satellite to come down because of its large areas of uninhabited water.[14] There has only been one recorded incident of a person being hit by human-made space debris. In 1997 an Oklahoma woman named Lottie Williams was hit in the shoulder by a 10 x 13 centimeters (5.1 in) piece of blackened, woven metallic material that was later confirmed to be part of the fuel tank of a Delta II rocket which had launched a U.S. Air Force satellite in 1996. She was not injured.

The USSR launched their nuclear-powered Cosmos 954 naval surveillance satellite on 18 September 1977; it disintegrated over northern Canada on 24 January 1978-possibly due to a collision with another object- resulting in the radioactive polluting of an area the size of Austria.

The USSR paid C$3 million to Canada on April 2, 1981, 'in full and final settlement of all matters connected with the disintegration of the Soviet satellite "Cosmos 954" in January 1978.' The Japanese Space Debris Study Group also disclosed that the rate of collision among pieces of space debris would increase about threefold by the year 2005, compared with 1987.

According to NASA's study, between 20,000-70,000 pieces of space debris were circling the earth, at an altitude of 800km-1000km; this space debris is out of control and non-function. There can be no doubt that as the number of space objects which are launched increases, the amount of debris re-entering the earth's atmosphere will also increase.

The possibility of a person being injured or property being damaged on the earth from the uncontrolled re-entry into the earth atmosphere of space objects is, however, slim. States Parties shall also take measures to avoid harming the environment of the earth through the introduction of extraterrestrial or other matter.

The largest space debris incident in history was the Chinese anti-satellite weapon test on January 11, 2007.

The intentional destruction of China's Fengyun-1C weather satellite via an anti-satellite (ASAT) device launched by the Chinese has created a mess of fragments fluttering through space.[15] China was widely condemned after their 2007 anti-satellite missile test, both for the military implications as well as the huge amount of debris it created. This is the largest single space debris incident in history in terms of new objects, estimated to have created more than 2,300 pieces (updated 13 December 2007) of trackable debris (approximately golf ball size or larger), over 35,000 pieces 1 cm (0.4 in) or larger, and 1 million pieces 1 mm (0.04 in) or larger.

Figure 4. The largest space debris incident in history was the Chinese Anti-satellite weapon test on January 11, 2007

Source: https://www.telegraph.co.uk/news/worldnews/1539948/Chinese-missile-destroys-satellite-in-space.html

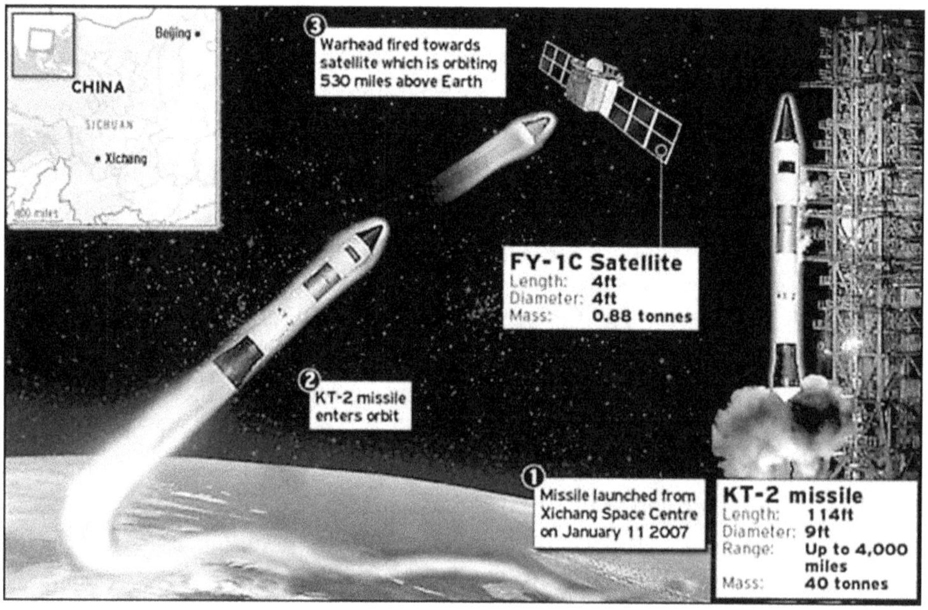

The test took place in the part of near Earth space most densely populated with satellites, as the target satellite orbited between 850 kilometer (530 mi) and 882 kilometers (548 mi). Since the atmospheric drag is quite low at that altitude, the debris will persist for decades. In June 2007, NASA's Terra environmental spacecraft was the first to perform a maneuver in order to prevent impacts from this debris. On 20 February 2008, the U.S. launched an SM-3 Missile from the USS Lake Erie specially to destroy a defective U.S. spy satellite thought to be carrying 1,000 pounds (450 kg) of toxic hydrazine fuel. Since this event occurred at about 250 km (155 mi) altitude, all of the resulting debris have a perigee of 250 km (155 mi) or lower. The missile was aimed to deliberately reduce the amount of debris as much as possible, and they had decayed by early 2008.

On 10 February 2009, at approximately 16:55 GMT, the inactive Russian Federation communications satellite "Cosmos 2251" and the operational United States satellite "Iridium 33" collided at an altitude of 790 km above the Earth.[16]

The collision created a cloud of nearly 700 pieces of space debris. Space debris remains in orbit for a considerable length of time and pose a risk to spacecraft orbiting Earth.

The Space Debris Mitigation Guidelines of the United Nations Committee on the Peaceful Uses of Outer Space (UNCOPUOS) adopted by UN General Assembly Resolution 62/217, aim to curtail the generation of such potentially harmful space debris. The debris cloud created by this collision is like a shotgun blast that threatens other satellites in the region.

Figure 5.
Source: http://news.bbc.co.uk/2/hi/7885051.stm

China has never been particularly apologetic about its contribution to the looming threat of space debris, but authorities might finally have to offer up some kind of conciliatory "*sorry we nearly bombed your village with huge chunks of used rocket.*"

16.3. ACTIVITIES OF THE UN COMMITTEE ON THE PEACEFUL USES OF OUTER SPACE

The UNCOPUOS, Legal Subcommittee was held in 58th Session, Vienna, Austria on April 1~11, 2019. Agenda Item 10 was General exchange of information and views on legal mechanisms relating to space debris mitigation and remediation measures, taking into account the work of the Scientific and Technical Subcommittee. Statement by the German Delegation was as the following.

The growing number of space debris unquestionably poses one of the biggest threats to the safe conduct of outer space activities. Research on the long-term evolution of space debris conducted by the German Aerospace Center (Deutsches Zentrum für Luft- und Raumfahrt; DLR) and ESA enables us to get a better understanding of the consequences of today's space activities for the future space environment.

The increasing numbers of small satellites in orbit as already observable today, as well as the announcements of large satellite constellations in Low Earth Orbit give reason for great concern. The observed trends make it imperative to strictly adhere to the Inter-Agency Space Debris Coordination Committee (IADC) and "*UNCOPUOS Space debris Mitigation Guidelines*" with an aim to limit the generation of new orbital debris by mitigating the risk of break-ups, and to conducting proper passivation and post-mission disposal procedures.

However, research reveals worrying low compliance rates with the space debris mitigation guidelines. It is crucial to raise awareness about proper debris mitigation standards and procedures as well as to find ways to improve compliance by both, governmental as well as private space actors.

It is appropriate to recall that any intentional destruction of an on-orbit spacecraft generating additional space debris poses a major safety threat to space activities conducted for the benefit and in the interest of all humankind and must therefore be avoided. Due to the energy converted during the impact of anti-satellite weapons even in low earth orbits any resulting space debris is uncontrollable and increases collision risks including in higher orbits.

Therefore, generally accepted international standards such as '*the Space Debris Mitigation Guidelines of COPUOS and the IADC* "as well as the "*Recommendations of the Group of Governmental Experts on Transparency and Confidence Building Measures in Outer Space*" activities urge responsible space actors to refrain from the intentional destruction of space objects.

Like already done in other forums, Germany has calls for a legally binding prohibition of the intentional destruction of space objects resulting in the generation of long-lasting debris, including in situations of armed conflict.[17]The UN Office for Outer Space Affairs (UNOOSA) and the European Space Agency (ESA) signed a joint statement on their wish to cooperate on the challenge of space debris in Vienna, Austria on May 24, 2019. .

The two organizations agreed to work together to increase global understanding and the consolidation of knowledge on space debris; to disseminate information on the latest research on space debris; to support the implementation of existing space debris mitigation guidelines; and to strengthen international cooperation and global awareness on space debris mitigation.

Space debris is widely defined as all non-functional, human-made objects, including fragments and elements thereof, in Earth orbit or re-entering into Earth's atmosphere. Space debris is increasing, in parallel with the growing number of objects launched into outer space, and pose exponential risks of collision, threatening space operations and limiting the development of a secure commercial space environment.

In 2018, the United Nations General Assembly expressed its deep concern about the fragility of the space environment and the impact of space debris, which is an issue of concern to all nations.

The United Nations Committee on the Peaceful Uses of Outer Space (COPUOS), of which UNOOSA is Secretariat, has paid particular attention to the issue of preventing and minimizing the creation of space debris.

Every year, States and organizations exchange information on their space debris research at the Committee's Scientific and Technical Subcommittee. One important result of those discussions has been a set of Space Debris Mitigations Guidelines, which were endorsed by the General Assembly in 2007. In addition, a compendium of space debris mitigation standards has been compiled by UNOOSA and, at the request of States, is made publicly available through UNOOSA's website.[18]

16.4. EUROPEAN CONFERENCE ON SPACE DEBRIS

The European conference on Space Debris Risks and Mitigation was held in Darmstadt, Germany on April 20, 2019. Since 1957, more than 4,900 space launches have led to an on-orbit population today of more than 18,000 tracked objects. Only 1,100 are functional spacecraft. The remaining 94% are space debris, i.e. objects which no longer serve any useful purpose. About 64% of the routinely tracked objects are fragments from some 250 breakups, explosions and collisions of satellites or rocket bodies. In addition, there is evidence of a much larger population of debris that cannot be tracked operationally.

An estimated number of 700,000 objects larger than 1 cm and 170 million objects larger than 1mm are expected to reside in earth orbits. Due to relative orbital velocities of up 56,000 km/h, centimeter-sized debris can seriously damage or disable an operational spacecraft, and collisions with object larger than 10 cm will lead to catastrophic break-ups, releasing hazardous debris clouds of which some fragments can cause further catastrophic collisions that may lead to an unstable debris environment in some orbit regions ("Kessler syndrome").

Space debris mitigation measures, if properly implemented by spacecraft designers and missions operators, can curtail the growth rate of the space debris population. Active removal, however, has been shown to be necessary to reverse the debris increase.[19]

The 7th European Conference on Space Debris was held on 18–21 April at ESA's Space Operations Centre in Darmstadt, Germany, was provided a unique forum for leading scientists, engineers, managers, space operators, industry, academia and policy-makers from all major spacefaring nations.[20]

During the 5th European Conference on Space Debris[21], 30 March to 2 April 2009, held at the European Space Agency, Darmstadt, Germany, experts from a wide spectrum of disciplines communicated their research results through 100 oral presentations and more than 40 poster presentations.

The conference was attended by some 330 participants from 21 countries, making this the largest dedicated space debris conference in the world. Key areas were measurements and debris environment characterization, environment modeling and forecasting (including orbit prediction aspects), risk analysis for the in-orbit and re-entry mission phases, protection and shielding, debris mitigation and re-mediation, and debris policies and guidelines. In the area of Space Surveillance, European researchers presented their proposals for optical and radar sensors, and for data processing methods for a European Space Situational Awareness System (SSA).

They looked at expected detection, correlation and cataloging performances. U.S. and Russian surveillance experts provided some insight into their experience gained over some 50 years of service. Many participants encouraged international cooperation and/or coordination of surveillance activities.

With the high quality presentations by a truly international community of experts, the 5th European Conference on Space Debris provided a comprehensive snapshot of the current state of space debris research.

16.5. THE INTER-AGENCY SPACE DEBRIS COORDINATION COMMITTEE

The Inter-Agency Space Debris Coordination Committee (IADC)[22] is an internati- onal forum of governmental bodies for the coordination of activities related to the issues of man-made and natural debris in space.

The Inter-Agency Space Debris Coordination Committee (IADC) is an inter-governmental forum whose aim is to co-ordinate efforts to deal with debris in orbit around the Earth founded in 1993.[23]

The primary purpose of the IADC is to exchange information on space debris research activities between member space agencies, to facilitate opportunities for co-operation in space debris research, to review the progress of ongoing co-operative activities and to identify debris mitigation options. [24]

One of its efforts is to recommend debris mitigation guidelines, with an emphasis on cost effectiveness, that can be considered during planning and design of spacecraft and launch vehicles in order to minimize or eliminate generation of debris during operations.

IADC is an international forum of national and international space agencies for the worldwide technical/scientific coordination of activities related to space debris in Earth orbit issues and provides technical recommendations.

The Inter-Agency Space Debris Coordination Committee (IADC) 13 member agencies include the following:

- ASI (Agenzia Spaziale Italiana)
- CNES (Centre National d'Etudes Spatiales)
- CNSA (China National Space Administration: 中国国家航天局)
- CSA (Canadian Space Agency)
- DLR (German Aerospace Center)
- ESA (European Space Agency)
- ISRO (Indian Space Research Organization)
- JAXA (Japan Aerospace Exploration Agency: 日本宇宙航空研究開発機構)
- KARI (Korea Aerospace Research Institute: 한국항공우주연구원)
- NASA (National Aeronautics and Space Administration)
- ROSCOSMOS (State Space Corporation, Russia)
- SSAU (State Space Agency of Ukraine[REMOVED TA FIELD])
- UK Space Agency (The United Kingdom)

A Steering Group and four specified Working Groups covering measurements (WG1), environment and database (WG2), protection (WG3) and mitigation (WG4) make up the IADC. This site provides information to the public about the IADC, its member agencies, and past and current debris-related activities. Furthermore, it is meant to act as a communication platform for the IADC member agencies. IADC members are national or international space and state organizations that carry out space activities through planning, designing, launching, or operating space objects.

IADC members should actively undertake space debris research activities and contribute to an increased understanding of space debris issues for the preservation of the orbital environment.

JAXA hosted the Annual Meeting in Tsukuba, Japan on June, 2018. ASI hosted the Annual Meeting in Rome, Italy (AS/HQ) on May 7~10, 2019. Though IADC annual meeting which will be hosted in Toulouse, France from 6 to 10 April 2020, but this annual meeting has decided to cancel the IADC 2020 conference due to the current outbreak of coronavirus (COVID-19). The 17th meeting of the Inter-Agency Space Debris Coordination Committee (IADC)[25] at Darmstadt, Germany discussed ways and methods to control the growing amount of orbiting debris. Radar and optical telescopes regularly track over 10,000 artificial objects in space. The number of untraceable objects in the size range from 1 cm to 10 cm, that could seriously damage an operational spacecraft, is estimated at between 100,000 and 150,000. The International Space Station (ISS) will be equipped with about 200 shields in order to defeat impacts of particulate up to about 1~2 cm size.

10.6. LEGAL PROBLEMS ON THE 1972 SPACE LIABILITY CONVENTION

The Liability Convention of 1972[26] provides that both intergovernmental organizations and State parties are liable on the basis of fault and non-fault for damage of their space objects, launch vehicles, or component parts thereof may cause in outer space.

The Liability Convention does not apply to damage caused to the launching State's nationals taking part in the launch (Article VII). It does not apply either to foreign nationals involved in the launching operations.

This exclusion confirms that the Liability Convention is especially set to protect *"innocent"* victims not taking part in this dangerous activity. The 1972 Liability Convention has been raised as a legal problems because there is no provision for compensation for damages caused by space accidents (space debris).

Finally, Article IX establishes that a claim for compensation for damage must be presented to a launching State through diplomatic channels. As to the identification of the moment in which the claim presented by the State has to be considered as unsatisfied by the launching State, the 1972 Convention does not contain any indication in this respect.

16.6.1. The Definition of Space Debris

Existing space national laws and international space treaties and agreements contain neither a definition nor a description of space debris. The seriousness of the debris problem for space operations, the possible confusion over the literal meaning of "debris," and the need to define the scope of debris all suggest the need for a legal term of art. According to the description of space object in the Liability Convention, all operational debris except litter appears to be "component parts," although jurists do not agree on this point.

16.6.2. The Domestic and International Legal Regime for Space Debris

Whether all fragmentation debris and micro-particulate matter are included is even more problematic, even if the broader interpretation of "space object" is invoked. Earlier law, including the international treaties and agreements on space, failed to address orbital debris explicitly. It is necessary for us to solve the question of jurisdiction for space debris and control over space debris, detection and identification of space debris, international responsibility for space debris, and possible remedies for damage caused by debris.

16.7. SPACE DEBRIS AND LIABILITY FOR DAMAGE CAUSED IN OUTER SPACE

The Liability Convention sets out a legal regime to provide compensation for damage caused in outer space by space objects. In outer space, liability is based on fault. It is significant to note that negotiations for the Liability Convention did not consider the question of the risks posed by space debris.

As a result, the negotiators did not address several liability issues of extreme importance related to damage caused by space debris. These issues include the meaning of *"damage"* and the reasonableness of a fault-based liability regime for damage caused in outer space by space debris. It would be possible to amend the Liability Convention so as to include damage to the outer space environment per se, based on the fact that outer space is a global commons.

The principle of fault-based liability is a further impediment to compensation for damage caused in outer space by orbital debris. Even if damage caused by space debris were within the scope of this regime, several other important legal issues, such as proof of negligence, gross negligence and contributory negligence, wilful-misconduct, among others, would remain unresolved.

16.7.1. Indirect Damage

Though the Liability Convention regulated the damage caused by the space objects, but it does not regulated the extent of a causal relationship between space activity and effect occurrence as well as the nature and scope of damage. Generally, It is adequate to compensate for the *"direct damage"* without cause of parameters as a cause of the space activity.

However, it was a hot controversy for *"indirect damage"* relating to the protection of victims in those day of drafting for this Convention. Reading the Article 1 of the Convention it is a clear that only direct *'damage',* and not indirect damage, is contemplated by the Liability Convention. During the preliminary discussion several delegates had pointed out that the question of indirect damage could cause great difficulties in practice.

As a hypothesis example for the theory of damages in these neighborhoods as a result of the satellite debris fallen caused the heart attack, or a businessman's object to the space physical results from damage to the businessman and the financial institutions to borrow money in damages.

It was discussed also the issues for the consolation money such as "mental damages" and "*profit loss*" in concluding with the Liability Convention. Especially some countries has not regulated the compensation for damages on the mental loss and has enacted the punitive damage for wrongdoer in their domestic law.

It was pointed out that the Liability Convention could not connected perfectly the protection of victims in the law of application.

16.7.2. Jurisdiction and Control for Space Debris

Who has jurisdiction and control over space debris? It is significant to note that negotiations for the Liability Convention did not consider the question of the risks posed by space debris. It is a desirable thing for us to amend some articles of the 1972 Liability Convention after studying more deeply the said five items of it.

16.7.3. Comment on the Draft for the International Instrument and Proposal on the New Legal Frame for the Prevention and Mitigation of Space Debris

16.7.3.1. Comment on the ILA's "Draft for the International Instrument of the Space Debris"

Prof. Dr. Karl Heinz Böckstiegel (Germany) and Prof Maureen Williams (Argentina) proposed a new *"Draft for the International Instrument on the Protection of the Environment from Damage Caused by Space Debris"*[27] at the Space Law Committee of International Law Association (ILA) with the result that the matter was discussed in depth among participants, as well as members of ILA 66th Buenos Aires Conference, Argentina held on August 16, 1994.

This new *"Draft for the International Instrument"* was passed by the Space Law Committee of the ILA same day. The final Draft has been under discussion and consultation by ILA members for fourteen years, at the ILA 67th Helsinki Conference in 1996, ILA 68th Taipei Conference in 1998, ILA 72nd Toronto Conference in 2006 and ILA 73rd Rio de Janeiro Conference in 2008[28] and ILA 74th Hague Conference in 2010.[29]

All the participants of the Buenos Aires Conference deliberated the contents of *"this final International Instrument"* at length, with the aim of protecting the interests of suffering States, damaged enterprises, and victims claiming for damage caused by launching States' space debris.[30]

It seems pertinent for this question of space debris to continue under review by the ILA Space Law Committee having in mind that it is now being discussed within UNCOPUOS. It is now important to encourage the examination of the legal aspects of space debris within the Legal Subcommittee of UN-COPUOS.

In pursuance of this target an important step forward was given in June 1996 and June 1797 by Prof. K. H. Böckstiegel who, as representative of the International Law Association, made a presentation to the Full Committee of UNCOPUOS at Vienna, Austria.

At the 1997 Session special reference was made to the work of the ILA on the *"Disputes Settlement related to Space Activities"* and to the Buenos Aires ILA Instrument on Space Debris. Following an invitation from the Former Director of the U. N. Office for Outer Space Affairs, Dr. Jasentuliyana, Prof. K. H. Böckstiegel submitted, in October 1997, a paper concerning the work of the ILA on space debris to be considered by the Scientific and Legal Subcommittee of UNCOPUOS during its Session (7-20 February 1998) at Vienna. In addition, many of Space Law Committee members have provided further comments on this matter and have made contributions tending to give more precision to some of the definitions embodied in the 1994 Buenos Aires International Instrument on Space Debris. Before summarizing these proposals, and in order to keep in line with the 1984 Paris Convention, it is suggested that the 1994 Instrument on Space Debris should, from now on, be referred to as the *"Draft for Convention on the Protection of the Earth Environment from Damage caused by Space Debris."*

Professor Carl. Q. Christol proposes the inclusion of a new subparagraph (d) in Article 1 to define "space object" which would consist of a *"man-made vehicle or entity launched into outer space."*

Professor Gabriella Venturini insists on the importance of including space debris on the agenda of the Legal Subcommittee of UNCOPUOS, noting that some delegations were already raising the issue within this body.

At the regional level, mention should be made of the XXI Meeting of Air and Space Law organized by the Latin American Association (ALADA) in Mexico, in May 1997.

One of the topics on the agenda was "*Legal Aspects of Space Debris*" and Professor M. A. Ferrer acted as Rapporteur.

I was commented some Articles of the new "*Draft for the International Instrument on the Protection of the Earth Environment from Damage Caused by Space Debris*" and presented my proposal on the new legal frame for the prevention and mitigation of space debris and the damage caused by the Space debris as the followings.

According to Articles 6, 7, and 8 in "*this Draft for the International Instrument*", it would appear that a basically faulty, unlimited liability system was adopted.

It would be almost impossible to prove the causes of accident and damage by space debris via the States, enterprises, legal entities and persons which suffered damage.

I would like to express my opinion that this "*International Instrument*" have to adopt an absolute and strict liability system in order to protect the victims caused by the damage due to the space debris accidents.

It is very difficult for victims to prove the cause of damage caused by space debris under faulty liability system.

It is desirable for us that this "*International Instrument*" would insert newly the word 'absolutely' after the word 'internationally' in Article 8, paragraph 2, in the final text of the International Instrument as the following Article.

Article 8 (International Liability), Each States or international organizations party to this Instrument that launches or procures the launching of a space object is internationally liable absolutely for damage arising there from to another State, persons or objects, or international organization party to this Instrument as a consequence of space debris produced by any such object.

According to the 1972 Liability, Article 2 states; 'A launching State shall be absolutely liable to pay compensation for damage caused by its space object on the surface of the earth or to aircraft in flight.' Under the regime of absolute liability, States will be liable under any circumstances except case of *force majeure*.

16.7.3.2. Space Debris and Liability Insurance

The space insurance will be keen to protect, indirectly, future missions that they underwrite by protecting the environment in which those assets will operate. Thus, a mission making a significant contribution to collision hazards amongst satellites and space debris in the orbit could be subjected to penalty premiums, related to the detrimental impact of the mission.

In the case of damage caused by deactivated satellites or space debris, upon payment of compensation for damage, the insurer must pay to the insured person the insurable amount thereof, in accordance with a general principle of insurance law. After Article 8, in the said International Instrument, I suggest the insertion of the following paragraph (*cf.* Annex 3): Article 8, (2).

The launching satellite State must cover liability insurance for damage caused by space debris in order to guarantee perfectly the compensation for personal and property damage of the damaged State, legal entity and person.

16.7.3.3. Draft for the Convention on the "Establishment of an International Organization for Protection of Earth Environment from Space Debris

I proposed also a new *"Draft for the Convention on the Establishment of an International Organization for Protection of Earth Environment from Space Debris (tentative title)"* under the United Nations Committee for the Peaceful Uses of Outer Space (UNCOPUOUS) in order to track, observe, detect and monitor the space debris as well as the prevention and mitigation of the damage caused by space debris.

It is a desirable and necessary for us that contracting states of the 1967 Space Treaty and the 1972 Liability Convention must establish an integrated monitoring system for the purpose of preventing and mitigating the damage caused by space debris in the legal frame as a special, affiliated organization under the auspices of the UNCOPUOS in order to protect the earth environment.

16.7.3.4. Draft for the Convention on the Establishment of an International Fund for Compensation of Space Debris' Damage

It is necessary and desirable for us to enact a new *"Draft for the Convention on the Establishment of an International Fund for Compensation for Space Debris Damage (tentative title)"* in order to protect the earth environment and space damaged States or space victim's. If the treaty was drafted and passed by the UNCOPUOUS General Assembly, the treaty member States are required to pay annual membership fees and funds according to the ratio of satellites and spacecraft launches and the ratio of national income of member States.

The funds accumulated in this international organization can compensate for the damages of the victims of the member states or victims who have suffered the unknown cause of space accidents, such as space debris, and can also be used for research expenses to remove or reduce space debris etc.

The international maritime law had been adopted already the *"International Convention on Civil Liability for Oil Pollution Damage of 1969"* and the *"International Convention on the Establishment of an International Fund for Compensation for Oil Pollution Damage of 1971"* such as a useful mechanism for ensuring the payment of compensation for oil pollution damage.

The aforementioned Special Space Agency of the UNCOPUOUS or a new International and Environmental Monitoring Organization should provide the funds needed to carry out such a survey efficiently, its purpose being the prevention of space debris accidents and the provision of compensation for damage caused by space debris.

I would also like to suggest the legislation of regulations with regard to the *"Draft for the Convention on the Establishment of an International Fund for Compensation for Space Debris Damage (tentative title)"*and under the auspices of the UNCOPUOUS and the establishment of the International Fund, as mentioned in Article 4 in the Rapporteur's final 'International Instrument.'

16.8. CONCLUSION

The obvious conclusion is that danger will become clearly occurrence posed by the uncontrolled re-entry of space debris into the earth's atmosphere. This danger is increased by the inability of modem space science and technology to predict the time of disintegration of non- functional and abandoned satellites and other space objects in outer space. What is important is that the potential for risk and damage is always present. Over the last few years, space debris has become a matter of increasing concern and has been identified as a serious threat to the further development of man's exploration and utilization of outer space.

The minimization of space debris requires policies of prevention and correction. In order to mitigate, if not eliminate, the presence of debris, these policies will have to be implemented by both active and passive measures. They should include the previously identified procedures and should be legal obligations.

The steady about the increase of space debris shares with many other pollution problems the characteristic that the broader public may well ignore it, until a serious incident which threatens or takes human life, or creates considerable economic damage, makes action essential. The space debris problem can only be effectively solved by international cooperation. Consultation and cooperation between space agencies are taking place with increasing frequency.

It is my firm opinion that only international and regional cooperation could solve the problem of environmental pollution, including damage caused by space debris meanwhile, States have to keep in mind that the exploration and use outer space is for the benefit and in the interests of all countries.

We should try to reach an international agreement, binding on all space-using nations, which encompasses the need to protect the space and earth environment. The Space Law Committee of ILA was sent it to the UNCOPUOS to recommend the legislation of the said new *Draft for the International Instrument*.

After the committee of space-faring nations has reached consensus on the basic steps to be taken, and has presented its findings to the Scientific and Technical Subcommittee of the UNCOPUOS, it may be appropriate for the Legal Subcommittee of UNCOPUOS to take up the matter, in order to debate and resolve the definition, jurisdiction and control, liability, and other legal issues that many legal experts consider need addressing in the preparation of an international agreement.

It is necessary for us to exchange the current and various information and view among the countries on the damage caused by space debris and to establish the integrated monitoring system in the world for the purpose of preventing and mitigating the damage caused by space debris as a special and affiliated organization under the United Nations. Furthermore it is necessary for us to amend some articles of the 1972 Liability Convention in order to protect victims in the case of the damage for compensations caused by the space accidents including space debris.

ADDITIONAL READING

Kim, D. H. (2008). *Essay for the Study of the International Air and Space Law*. Korea Studies International Co. Ltd.

Space Debris Group of Japanese Society for Aeronautical and Space Sciences. (1993). *Report of Space Debris Group*. Author.

The International Law Association. (1994). Report of the Sixty-Sixth Buenos Aires Conference. Author.

The International Law Association. (2010). *Space Debris.* Report of the ILA 74th Hague Conference, Space Law Committee.

UN Committee on the Peaceful Uses of Outer Space, Scientific and Technical Subcommittee. (1994). 31st Session, Vienna, 21 February-4 March 1994, Agenda item 8. Space Debris Report of the International Astronautical Federation, UN Doc. A/ AC.105/570. 25 February 1994, p.1.

KEY TERMS AND DEFINITIONS

CNSA: China National Space Administration (CNSA) (国家航天局) is the national space agency of China. It is responsible for the national space program[2] and for planning and development of space activities. CNSA and China Aerospace Science and Technology Corporation (CASC) assumed the authority over space development efforts previously held by the Ministry of Aerospace Industry. It is a subordinate agency of the State Administration for Science, Technology and Industry for National Defence (SASTIND), itself a subordinate agency of the Ministry of Industry and Information Technology (MIIT).

ISRO: Indian Space Research Organization (ISRO) is the national space agency of the Republic of India, headquartered in Bengaluru. It operates under Department of Space (DOS) which is directly overseen by the prime minister of India while chairman of ISRO acts as executive of DOS as well. ISRO is the primary agency in India to perform tasks related to space based applications, space exploration and development of related technologies.

JAXA: The Japan Aerospace Exploration Agency (JAXA) (国立研究開発法人宇宙航空研究開発機構, *Kokuritsu-kenkyū-kaihatsu-hōjin Uchū Kōkū Kenkyū Kaihatsu Kikō*, literally "National Research and Development Agency on Aerospace Research and Development") is the Japanese national aerospace and space agency. Through the merger of three previously independent organizations, JAXA was formed on 1 October 2003. JAXA is responsible for research, technology development and launch of satellites into orbit, and is involved in many more advanced missions such as asteroid exploration and possible human exploration of the Moon. Its motto is *One JAXA* and its corporate slogan is *Explore to Realize* (formerly *Reaching for the skies, exploring space*).

KARI: The Korea Aerospace Research Institute (KARI: 한국항공우주연구원) established in 1989, is the aeronautics and space agency of South Korea. Its main laboratories are located in Daejeon, in the Daedeok Science Town. KARI's vision is to continue building upon indigenous launch capabilities, strengthen national safety and public service. The Korean Lunar Exploration Program (KLEP) is divided in two phases. Phase 1 incorporates the launch and operation of a lunar orbiter called Korea Pathfinder Lunar Orbiter (KPLO), which will be the first lunar probe by South Korea, meant to develop and enhance South Korea's technological capabilities, as well as map natural resources from orbit. Phase 2 will include a lunar obiter, a lunar lander, and a rover to be launched together on a KSLV-II South Korean rocket from the Naro Space Center, by 2030.

Naro (나로) Space Center: Naro Space Center is a South Korean spaceport in South Jeolla's Goheung County, operated by the state-run Korea Aerospace Research Institute. The spaceport is located about 485 km from south of Seoul, Korea. It supported 4 launches and will also support the KSLV-II launch in 2021, and SSLV launches in 2025.

US Space Surveillance Network (SSN): The United States Space Surveillance Network detects, tracks, catalogs and identifies artificial objects orbiting Earth, e.g. active/inactive satellites, spent rocket bodies, or fragmentation debris. The system is the responsibility of the Joint Functional Component Command for Space, part of the United States Space Force (formerly USSPACECOM (United States Space Command).

ENDNOTES

1 *Report (Proceeding)* of the 66th Conference of the International law Association (ILA) at Buenos Aires, Argentine on August 14~20, *Space Law,* Space Law Committee, pp. 317-325.
2 https://www.voanews.com/english/news/a-13-2008-04-08-voa28-66747082.html
3 https://en.wikipedia.org/wiki/Naro_Space_Center
4 UN Committee on the Peaceful Uses of Outer Space, Scientific and Technical Subcommittee, 31st Session, Vienna, 21 February-4 March 1994, Agenda item 8, *Space Debris Report of the International Astronautical Federation*, UN Doc. A/ AC.105/570. 25 February 1994, 1.
5 UN Committee on the Peaceful Uses of Outer Space, Scientific and Technical Subcommittee, 31st Session, Vienna, 21 February-4 March 1994, Agenda item 8, *Space Debris Report of the International Astronautical Federation*, UN Doc. A/ AC.105/570. 25 February 1994, 1.
6 Space Debris Group of Japanese Society for Aeronautical and Space Sciences, '*Report of Space Debris Group of 1993*' at 1-12.
7 Space Debris Study Group of Japanese Society for Aeronautical and Space Science, *supra*, 1.
8 John H. Carver, '*Space Debris*', First European Conference on Space Debris by European Space Agency Darmstadt, Germany, 5-7 April 1993 (speech paper), at 3.
9 United Nations, General Assembly, A/AC. 105/570, 25 February 1994, Scientific and Technical Sub-committee, 31st Session, Vienna, 21 February- 4 March, 1994, Agenda item 8, *Space Debris, Report of the International Astronautical Federation,* 1.
10 Ray A Williamson & Richard Obermann, *New Challenges in International Orbital Debris Policy*, 45th Congress of the International Astronautical Federation, 9-14 October 1994, Jerusalem, Israel, at 1.
11 P.T.I., Narendra Modi Announces Success of Mission Shakti, India's Anti-Satellite Missile Capability, THE HIND (Mar. 27, 2019), https://www.thehindu.com/news/national/narendra-modi-announces-success-of-mission-shakti-indias-anti-satellite-missile-capability/article26651731.ece [https://perma.cc/LLJ8-8WGL].
12 *See, e.g.*, Helen Regan, *India Anti-Satellite Missile Test a 'Terrible Thing,' NASA Chief Says*, CNN (Apr. 2, 2019), https://www.cnn.com/2019/04/02/india/nasa-India-anti-missile- tes-intl/index.html [https://perma.cc/ND29-RCK3].
13 https://en.wikipedia.org/wiki/Space_debris
14 https://en.wikipedia.org/wiki/Space_debris
15 http://www.space.com/3415-china-anti-satellite-test-worrisome-debris-cloud-circles-earth.html
16 http://celestrak.com/events/collision.asp
17 https://wien-io.diplo.de/iow-en/news/statement-debris/2208724
18 https://www.unoosa.org/oosa/en/informationfor/media/2019-unis-os-510.html
19 https://c-ssystems.de/european-conference-space-debris-and-mitigation
20 https://www.esa.int/Safety_Security/Space_Debris/European_conference_on_space_debris_risk_and_mitigation

21 http://www.esa.int/esaCP/SEM3C6KTYRF_index_0.html

22 https://www.iadc-home.org/

23 https://en.wikipedia.org/wiki/Inter-Agency_Space_Debris_Coordination_Committee

24 https://www.iadc-home.org/what_iadc

25 Doo Hwan Kim, *Essay for the Study of the International Air and Space Law,* Korea Studies International Co. Ltd (2008), at 421.

26 http://www.oosa.unvienna.org/oosa/en/SpaceLaw/gares/html/gares_26_2777.html

27 The International Law Association, "*Report of Space Law Committee, ILA 66th Buenos Aires Conference of 1994*", at 317-324.

28 https://www.ila-hq.org/en/committees/index.cfm/cid/29269.

29 Report (Proceeding) of the ILA 74th Hague Conference in 2010, *Space Law*, Space Law Committee, at 269.

30 The International Law Association, "*Report of the Sixty-Sixth Buenos Aires Conference of 1994*", at 305-317.

Chapter 17
Proposal of Establishing an Asian Space Agency

ABSTRACT

This chapter describes the author's proposal on the necessity of Establishing an Asian Space Agency (ASA). The establishment of the ASA will promote the international cooperation among Asian countries in space exploitation, research and technology, as well as their space application and developments, much like the European Space Agency in Paris. It is desirable and necessary for us to establish ASA in order to promote cooperation in space policy, law, science, technology, and industry among Asian countries. The creation of an ASA would lead to a strengthening of the cooperation deemed essential by the Asian community towards joint undertakings in space and would act as a catalyst for the efforts on space exploitation and allow resources, technology, manpower, and finances to be centrally managed in an independent fashion to the benefit of Asian countries. It is desirable and necessary for us to establish the ASA in order to develop the space industry to strengthen friendly relations and to promote research cooperation among Asian countries.

17.1. NECESSITY OF ESTABLISHING AN ASIAN SPACE AGENCY

The idea of creating an Asian Space Agency (hereinafter referred to ASA) is only my academic and practical opinion. The creation of an ASA would lead to a strengthening of the cooperation deemed essential by the Asian community towards joint undertakings in space and would act as a catalyst for the efforts on space exploitation and allow resources, technology, manpower and finances to be centrally managed in an independent fashion to the benefit of Asian countries.

"Space already exists for Asia, so the question is what do we do there." It's our job to make sure that all opportunities are used to integrate the power of space exploitation among the Asian countries. In the 21st century, space science and technology will develop with ever greater rapidity.

Having developed rapidly for about half a century, some Asian developed countries with space activities have scored remarkable achievements that greatly promoted the development of social productivity and progress. The continuous development and application of space technology has become an important

DOI: 10.4018/978-1-7998-7407-2.ch017

role in the modernization drive of Asian countries. The emergence of aerospace technology in Asian countries has brought huge contributions to economic and social progress.

Today, the Asian aerospace industry has become one of the world's most vigorous and promising high-tech industries. But in some Asian countries accidents have occurred over the past quarter of a century, owing to failure of satellite launches etc.

If the ASA will be established to promote cooperation among Asian countries in the near future, I believe that we could mitigate launching accidents by prevention, through analyzing the causes of accidents and cooperating jointly with the excellent space technicians in Asian countries.

Though some Asian space scientists already have a good multinational program, sometimes they have to rely on the Americans, Europeans or Russians to launch their space experiments or small satellites for them. NASA's (US National Aeronautics and Space Administration) [1], RASA's (Russian Aeronautics and Space Agency: RASA)[2] and ESA's (European Space Agency) [3] programs for launchers, science, telecommunications, earth observation, remote sensing, international space station and manned space flight have demonstrated great competence.

Space ventures and their applications have become normal everyday activities in the global community. The ASA is not only a part of Asia's legal, economic and social fabric, but also an organization of technical cooperation in Asian countries.

To sustain this success in the 21st Century, Asian countries must continue to consider establishing the ASA for opportune and carefully conceived multinational space projects and for help in improving their industrial competitiveness.

It is desirable and necessary for us to establish ASA in order to promote cooperation in space policy, law, science, technology and industry among Asian countries.

If the head of the Asian countries would agree to establish ASA at a summit conference, I am sure that it is possible to establish ASA in the near future.

The establishment of the ASA will promote the international cooperation among Asian countries in space exploitation, research and technology, as well as their space application and developments, much like the European Space Agency in Paris.

It is very important that ministers, high ranking officials, staff members of space agencies or authorities, space law professors, lawyers, scientists arid technicians from Asian countries agree in advance on the establishment of the ASA.

17.2. HISTORY OF THE REGIONAL AND INTERNATIONAL COOPERATION IN THE ASIAN COUNTRIES

Especially among Asian countries situated in the same geographic area, motivations of a political, economic, technical and social nature have strongly promoted cooperation in the field of space communications and industry as well as in political and economic organizations.

As a result, several regional, political and economic organizations were set up, such as Asia-Pacific Economic Cooperation (APEC) [4]), the Association of South East Asian Nations (ASEAN)[5]), the Asia-Europe Meeting (ASEM) [6]) and cooperative conferences such as the "Asia-Pacific Regional Space Agency Forum" at Tokyo, Japan, the "Ministerial Conference on Space Applications for Sustainable Development in Asia and the Pacific", the "Asian-Pacific Multilateral Space Technology Cooperation

Symposium" at Beijing, China, and the "UN-ESCAP Ministerial Conference on Space Applications for Sustainable Development in Asia and the Pacific" in India.

17.2.1. The Republic of Korea

The aerospace industry may create many benefits for the 21st century with new materials and life science industries. KARI strives to strengthen the international cooperation in the space arena by expanding its participation in international governmental and non-governmental organizations in order to keep abreast of global trends in space development.

Also, as part of efforts to contribute to the international space community, KARI has been offering a 2-week course of KARI International Space Training (KARIST) for space experts in particular from developing countries since 2010.

The main goal of this training program is to provide various opportunities for the beginning professionals to develop their space capacities and to serve as a preliminary platform to promote international cooperation among participants or with Korea.

KARI will continue to strengthen the global partnership with other countries and international organizations by actively engaging in international projects such as space exploration, resolving global issues and so on, all of which are key parts of our policy to facilitate space diplomacy to strengthen Korea's status in the international community.

KARI has solidified its infrastructure, and has made a great effort in research and development. KARI will do its utmost to securely establish the aerospace industry by the year 2000, and eventually develop into a world-class aerospace research center.[7])

In order to digest and introduce state of the art aerospace technologies, KARI has signed technical agreements with 17 organizations in the United States, England, France, Russia, China, Israel, and Poland. In the future, KARI will enlarge technical cooperation with aerospace-related research institutes in developing countries such as Brazil etc.

17.2.2. Japan

Japan led the establishment of the Asia-Pacific Regional Space Agency Forum (APRSAF).

The APRSAF holds a regular meeting on international cooperation in the Asian Pacific region and 26 countries from the region participate. 4The Asia-Pacific Regional Space Agency Forum (APRSAF), sponsored by NASDA and JAXA, is a conference to exchange views and perspectives on space development in the Asian region.

This forum has been held in 1994, 1995, 1997, 1999, 2000,[8], 2001, 2003, 2004, 2005, 2006, 2007 since 1993. As of January 2007, thirteen meetings have been held in Jakarta, Indonesia. This APRSAF also has been held every year from 2008 to 2019 in the Asia-Pacific Region.

The 25th Session of the Asia-Pacific Regional Space Agency Forum (APRSAF-25), Singapore, November 6-9, 2018. The 26th Session of the Asia-Pacific Regional Space Agency Forum (APRSAF-26), Nagoya, Japan, November 26-29, 2019.

The 27th Session of the Asia-Pacific Regional Space Agency Forum (APRSAF-27), Hanoi, Vietnam, October 27-30, 2020.[9]

The co-host countries so far have been Australia, India, Indonesia Malaysia, Mongolia, Philippines, Singapore, the Republic of Korea, Thailand, Vietnam and etc. The first four and the sixth session were convened at Tokyo, Japan.

The fifth session was convened at Ulaanbaatar, Mongolia. The Seventh Forum was held at the Institute of Industrial Science of the University of Tokyo from June 19-22, 2000. The three-day Forum ended successfully with enthusiastic discussions and exchanges of opinions among about 120 participants from sixteen Asia-Pacific countries and four international organizations. 438) NASDA teamed up with the Science and Technology Agency of Japan (STA) and the Institute of Space and Astronautical Science (ISAS), to hold the Seventh Session of the Asia-Pacific Regional Space Agency Forum (APRSAF-7). The Eleventh Session of APRSAF (APRSAF-11) was held from 3rd to 5th November 2004 in Canberra, Australia, jointly organized by the Australian Cooperative Research Centre for Satellite Systems (CRCSS), MEXT and JAXA. Under the main theme entitled "Toward Expansion of the Space Community", more than 100 local, Asia-Pacific and overseas participants took part in this Forum.

The Twelfth Session of APRSAF (APRSAF-12) was held on October, 2005 at Kitakyushu in Japan, organized by JAXA. APRSAF-12 also discussed how to develop human resources of the next generation in terms of contributing to the enhancement of space capability in this region.

The Thirteenth Session of APRSAF (APRSAF-13) was held from December 5- 7, 2006 in Jakarta, Indonesia, jointly organized and co-sponsored by the Ministry of Research and Technology Indonesia (RISTEK), National Institute of Aeronautics and Space (LAPAN), the Japanese Ministry of Education, Culture, Sports, Science and Technology (MEXT) and Japan Aerospace Exploration Agency (JAXA).[10])

Figure 1. The 26th Session of the Asia-Pacific Regional Space Agency Forum (APRSAF-26), Nagoya, Japan, November 26-29, 2019.

The APRSAF has provided an opportunity for discussing possibilities of future cooperation among space agencies in the Asia-Pacific region, and the participants included leading space industries from Asia-Pacific nations. The Participating Countries and Organizations were: Bangladesh, Canada, China,

India, Indonesia, Japan, Korea, Malaysia, Mongolia, Nepal, New Zealand, Pakistan, Kiribati, Russia, Singapore, Sri Lanka, Thailand, Vietnam, United States, UN Office for Outer Space Affairs (OOSA), Economic and Social Commission for Asia and the Pacific (ESCAP), and the International Space University.

NASDA's Bangkok Office has been promoting the relationship between Japan and other Asian countries in the space field, especially in the field of research and application of remote sensing technology from space.

In the Asian region there are many advantages in using remote sensing data for the purpose of natural disaster monitoring, environment monitoring, map generation etc. NASDA has been promoting the utilization of Japanese satellite data and has conducted some cooperative projects, seminars, training courses etc. in the Asian region.

In 2006, the "Sentinel-Asia (Asian supervisors)" was inaugurated to construct the "Asian disaster preparation and risk management system," comprising of 34 organizations from 18 countries and four international organizations.

In April, 2005, JAXA proposed its long-term vision up to 2025 for the field of aerospace. JAXA has started activities achieve this goal. JAXA puts emphasis on international cooperation, and currently most of the projects have been conducted through international cooperation.

As part of JAXA's cooperative relations with Asia-Pacific countries, JAXA mutually receive data directly from earth observation satellites and conduct cooperative research using observation data with Thailand, Australia, Indonesia, China, Malaysia, and Korea. With the support of the Kiribati government, since 1976 NASDA has established a downrange station on Christmas Island. The station is currently maintained by JAXA. In 2006, JAXA and the Korean Aerospace Research Institute (KARI) signed the "Memorandum of Agreement in Aerospace Fields."

17.2.3. China

The China National Space Administration (CNSA) was established in 1993, as a government institution to develop and fulfill China's due international obligations, with the approval by the Eighth National People's Congress of China (NPC). The Ninth NPC assigned CNSA[11)] as an internal structure of the Commission of Science, Technology and Industry for National Defense (COSTIND).[12]In 1994, together with the Economic and Social Commission for Asia and the Pacific (ESCAP)[13], China hosted in Beijing the first Asian-Pacific regional *"Ministerial Conference on Space Applications for Sustainable Development in Asia and the Pacific,"* and the "*Beijing Declaration*" issued after the conference has had a far-reaching influence.

In September 1999, in collaboration with the UN and European Space Agency(ESA), the Chinese government held in Beijing the *"Symposium on Promoting Sustainable with Space Applications."*

From July to August 2000, together with the United Nations Office for Outer Space Affairs (UNOOSA) and ESCAP, relevant departments of the Chinese government opened the Short-term Training Course *for Asia-Pacific Multilateral Cooperation in Space Technology and Applications.*

Trainees from ten developing countries in the Asia-Pacific region attended the course. China attaches great importance to space cooperation in the Asia-Pacific region. In 1992, China, Thailand, Pakistan and some other countries jointly sponsored the "*Asian-Pacific Multilateral Space Technology Cooperation Symposium.*" [14)]

According to the impetus of such regional cooperation, the governments of China, Iran, the Republic of Korea, Mongolia, Pakistan and Thailand signed the *"Memorandum of Understanding on Cooperation in Small Multi-Mission Satellite and Related Activities"* in Thailand in April 1998.

Besides the signatory countries, other countries in the Asia-Pacific region may also join the cooperative project, which has helped to enhance the progress of space technology and space application in the Asia-Pacific region. Since 1988, China has provided other developing countries every year with scholarships for long-term space technology training.

From 2001 to 2005, China's space industry has developed rapidly, making many achievements.

A group of research and development and testing bases of the advanced world level has been built, and the system of research, design, production and testing has been further improved, markedly enhancing the country's basic capabilities in space science and technology. With breakthroughs in important key technologies, the overall level of China's space technology has been improved remarkably.

Having made a historic break-through in manned space flight, China has embarked on a comprehensive lunar exploration project. Space application systems have taken shape, the range of application has been further expanded, application benefits have been noticeably enhanced, and important achievements have been made in space scientific experiments and research in this regard.

Over the past five years, China has independently developed and launched 22 different types of man-made satellites, upgrading its overall level in this field markedly. Over the past five years, "Long March (長征)" rockets independently developed by China have made 24 consecutive successful flights, and their major technological functions and reliability have been notably upgraded. From October 1996 to the end of 2005, "Long March" rockets made 46 consecutive successful flights.

The construction of four spacecraft launching sites at Jaquan (酒泉), Xichang (西昌), Taiyuan (太原) and Wenchang(文昌) has made new progress, and their comprehensive test and launch capabilities have been enhanced. On November 20 and 21, 1999, China launched and retrieved the first "Shenzhou (神舟)" unmanned experimental spacecraft.

It then launched three more "Shenzhou" unmanned experimental spacecraft not long afterwards. On October 15 and 16,2003, it launched and retrieved the "Shenzhou V" manned spacecraft, China's first of its kind.

Having mastered the basic technologies for manned spacecraft, China became the third country in the world to develop manned space flight independently. From October 12 to 17, 2005, the "Shenzhou VI" manned spacecraft completed a five-day flight with two astronauts on board. This was the first time for China to have men engage in experiments in space, another major achievement in the sphere of manned space flight.

The Shenzhou 7 was the third human space flight of the Chinese space program. The mission, which included an extra-vehicular activity (EVA) carried out by crew members Mr. Zhai Zhigang and Mr. Liu Boming marked the commencement of the second phase of the Chinese government's Project 921.

The Shenzhou 7 spacecraft carrying the three crew members was launched September 25, 2008 by a Long March 2F rocket which lifted off from the Jiuquan (酒泉) Satellite Launch Center. The mission lasted three days, after which the craft landed safely in Siziwang Banner central Inner Mongolia September 28, 2008 at 17:37 CST. The EVA carried out during the flight makes China the third country to have conducted an EVA, after Russia and the United States.

The Shenzhou 7 spacecraft for the first Chinese space mission to carry a three-person crew for several days and conduct a full operation. On September 27, 2008 Zhai Zhigang wearing a Chinese-developed Feitian space suit, conducted a 20-minute spacewalk, the first ever for a Chinese astronaut[15)]

Shenzhou 10 (Chinese: 神舟10号) was a crewed spaceflight of China's Shenzhou program that was launched on 11 June 2013. It was China's fifth crewed space mission. The mission had a crew of three astronauts. The Shenzhou spacecraft docked with the Tiangong-1 trial space laboratory module on 13 June, and the astronauts performed physical, technological, and scientific experiments while on board. Shenzhou 10 was the final mission to Tiangong 1 in this portion of the Tiangong program. On 26 June 2013, after a series of successful docking tests, Shenzhou 10 returned to Earth[16]

Shenzhou 11 was a crewed spaceflight of the Shenzhou program of China, launched on 17 October 2016 from the Jiuquan Satellite Launch Center. It was China's sixth crewed space mission, and its longest to date, at 33 days. Two days after launch, it docked with the Tiangong-2 space laboratory, which had been launched on September 15, 2016.[17]

Shenzhou 12 is a Chinese spaceflight planned to launch around 2021. The flight will mark the seventh crewed Chinese spaceflight and the seventh crewed flight of the Shenzhou programme. The spacecraft is scheduled to carry two to three CNSA Taikonauts (Astronaut) on the first flight to Tianhe-1 (天和 1), the first module of the Chinese large modular space station, scheduled for launch on a Long March 5B rocket sometime in 2021.[18]

Following the first test launch of a Long March 5B rocket on 5 May 2020, it was confirmed that the flight crew had been internally named, although no details about the Taikonauts who will fly the mission have been made public yet. It was also at one point speculated that the mission might involve one or more female Taikonauts, and could mark the first time a Chinese woman commands a space mission.

"The Outline of the 11th Five-Year Program for National Economic and Social Development" and "The National Guideline for Medium-and Long-term Plans for Science and Technology Development (2006-2020)" formulated by the Chinese government in 2006 put the space industry in an important position.

Over the past five years, China has developed bilateral space cooperation with a host of countries.

It has successively signed 16 international space cooperation agreements and memorandums with13 countries, space agencies and international organizations, and propelled multilateral cooperation in space technology and its application in the Asia-Pacific region and the process of establishing a space cooperation institution for the region.

China has joined relevant activities sponsored by the United Nations and other relevant international organizations, and supported international space commercial activities.

These measures have yielded positive results.

Over the past five years, China has signed cooperation agreements on the peaceful use of outer space and space project cooperation agreements with Argentina, Brazil, Canada, France, Malaysia, Pakistan, Russia, Ukraine[REMOVED TA FIELD], the ESA and the European Commission, and has established space cooperation sub-committee committee or joint commission mechanisms with Brazil, France, Russia and Ukraine.

It has signed space cooperation memorandums with space organizations of India and Britain, and has conducted exchanges with space-related bodies of Algeria, Chile, Germany, Italy, Japan, Peru and the United States. China continues to collaborate with Brazil on the Earth resources satellite program.

China and France have developed extensive space exchanges and cooperation. The space cooperation between China and Russia has produced marked results. Within the framework of the Space Cooperation Sub-Committee of the Committee for the Regular Sino-Russian Premiers' Meeting, a long-term cooperation plan has been determined. China has unfolded space exchanges and cooperation with Ukraine[REMOVED TA FIELD].

China and the ESA have carried out the Sino-ESA Double Star Satellite Exploration of the Earth's Space Plan. China's relevant departments and the ESA have implemented the "Dragon Program," involving cooperation in Earth observation satellites, having so far conducted 16 remote-sensing application projects in the fields of agriculture, forestry, water conservancy, meteorology, oceanography and disasters.

In October 2005, the representatives of China, Bangladesh, Indonesia, Iran, Mongolia, Pakistan, Peru and Thailand signed the Asia-Pacific Space Cooperation Organization (APSCO) Convention in Beijing, and in June 2006 Turkey signed the Convention as well. APSCO will be headquartered in Beijing.[19]

This marks a significant step toward the official establishment of APSCO. China continues to promote the Asia-Pacific Region Multilateral Cooperation in Small Multi-Mission Satellites Project. Together with Bangladesh, Iran, the Republic of Korea, Mongolia, Pakistan and Thailand, China has started the joint research, manufacture and application of small multi-mission satellites, to be launched in 2007.[20)] The Third Summer Camp of APSCO Student Small Satellite (SSS) Project organized by APSCO and hosted by Shanghai Jiao Tong University (SJTU) was officially kicked-off on July 15, 2019 in Shanghai, China. Over 40 participants from APSCO Member States have participated the program.

Figure 2. The Final Acceptance Review Meeting on Data Sharing Service Platform (DSSP) Phase II of APSCO was held in Beijing, China on Dec. 17-20

The previous two Summer Camps of the SSS Project, which is a satellite constellation under manufacturing by universities from Member States with the guidance of space agencies, were successfully organized by APSCO and achieved fruitful outcomes in 2017 and 2018, respectively.

The Final Acceptance Review Meeting on Data Sharing Service Platform (DSSP) Phase II of APSCO was held during December 17-20, 2019 at APSCO Headquarters in Beijing, China. The opening session was addressed by Dr. Li Xinjun, Secretary-General of APSCO. He welcomed all participants and thanked for their participation in the meeting. He emphasized on the importance of the project, introduced latest situation of DSSP pilot projects and encouraged experts to have a full discussion and comments-sharing for better operating of DSSP Phase II.

The meeting was chaired by Mr. Xu Yansong, Director General of the Department of Program Operation and Data Service of APSCO. "Test Plan and Test Instructions, Test Report, Project Development Summary Report, User Manual, Administrator Manual" were reviewed by experts. Delegates from Argentina National Space Activity Commission (CONAE) observed the meeting.

China takes a positive part in activities organized by the United Nations Committee on the Peaceful Uses of Outer Space (UNCOPUOS) and its Scientific and Technical Subcommittee and Legal Subcommittee.

17.2.4. India

The national space program was formally organized in June 1972 with the setting up of the Space Commission and the Department of Space (DOS) and the Indian Space Research Organization (ISRO) to promote the development and application of space technology and space science for the economic benefit of the nation.[21)] The prime objective of ISRO has been to develop space technology and its application to various national tasks.

Since 1969, when it was set up, ISRO has established space systems like the INSAT for telecommunication, television broadcasting and meteorological services, and the Indian Remote Sensing Satellites (IRS) for resources monitoring and management.

ISRO has also developed the satellite launch vehicles PSLV and GSLV to place these satellites in the required orbits.[22)] The overall coordination of the space program is carried out by program offices of ISRO headquarters in different areas like satellite communication, earth observation systems, launch vehicle program, space science, technology transfer and industry coordination, international cooperation, publications and public relations and, budget and economic analysis.

DOS implements the policies framed by the Space Commission. Research and development activities are carried out through ISRO, the National Remote Sensing Agency (NRSA), the Physical Research Laboratory (PRL), the National Mesosphere-Stratosphere-Troposphere Radar Facility (NMRF) and other agencies.

An Advisory Committee on Space Sciences (ADCOS) guides the research programs in the area of space science.

The Antrix Corporation Limited, Bangalore, established in 1992, is a wholly-owned Government of India company for commercial marketing of space products and services.

The Secretariat of DOS and the Headquarters of ISRO are located at Bangalore.

The International cooperation was involved with the establishment of TERLS, the conduct of SITE & STEP, the launches of Aryabhata, Bhaskara, APPLE, IRSIA and IRS-IB / satellites manned space mission, etc.

India has established a Center for Space Science and Technology Education in Asia and the Pacific (CSSTE-AP) that is sponsored by the United Nations. Lastly, India hosted the Second UN-ESCAP Ministerial Conference on Space Applications for Sustainable Development in Asia and the Pacific in November 1999.

International cooperation is a strategic area for a space programme because relationships with other countries are influenced by political, economic, cultural, and human personality factors as well as scientific and technological factors. India has always recognized that space has dimension beyond national considerations, which can only be addressed along with international partners.

Over the years, as ISRO has matured in experience and technological capabilities, the scope for cooperation has become multifaceted.

While exploratory missions beyond the earth are the natural candidates for such cooperative efforts, there are many other themes like climate change impacts on earth, space science and planetary exploration that are of interest to international cooperation because of their global impact. International cooperation has been part of Indian space programme since inception.

Establishment of Thumba Equatorial Rocket Launching Station (TERLS), conduct of Satellite Instructional Television Experiment (SITE) and Satellite Telecommunication Experiment Project (STEP), launches of Aryabhata, Bhaskara, Ariane Passenger Payload Experiment (APPLE), IRS-IA, IRS-IB satellites, INSAT series of satellites, Mission to Moon, etc., have the components of international cooperation.

ISRO is pursuing bilateral and multilateral relations with space agencies and space related bodies with the aim of building and strengthening existing ties between countries; taking up new scientific and technological challenges; refining space policies and defining international frameworks for exploitation and utilization of outer space for peaceful purposes. Internationally India is viewed by space faring nations as an emerging space power, capable of achieving its goals in a more cost effective and time-efficient manner. Specifically, the developing countries look to India for assistance in building up their capabilities to derive benefits of space technology.

The scope of international cooperation has become wider and diverse, as ISRO has made tremendous progress in recent time.

Formal cooperative arrangements in the form of either Agreements or Memoranda of Understanding (MoU) or Framework Agreements have been signed with Afghanistan, Algeria, Argentina, Armenia, Australia, Bahrain, Bangladesh, Bolivia, Brazil, Brunei Darussalam, Bulgaria, Canada, Chile, China, Egypt, Finland, France, Germany, Hungary, Indonesia, Israel, Italy, Japan, Kazakhstan, Kuwait, Maldives, Mauritius, Mexico, Mongolia, Morocco, Myanmar, Norway, Peru, Portugal, Republic of Korea, Russia, Sao-Tome & Principe, Saudi Arabia, Singapore, South Africa, Spain, Sultanate of Oman, Sweden, Syria, Tajikistan, Thailand, The Netherlands, Tunisia, Ukraine[REMOVED TA FIELD], United Arab Emirates, United Kingdom, United States of America, Uzbekistan, Venezuela, Vietnam. [23]

Also formal cooperative instruments have been signed with international multilateral bodies like European Centre for Medium Range Weather Forecasts (ECMWF), European Commission, European Organization for the Exploitation of Meteorological Satellites (EUMETSAT), European Space Agency (ESA) and South Asian Association for Regional Cooperation (SAARC).

India continues to play active role in deliberation on Scientific and Technical and Legal sub-committees of the United Nations Committee on the Peaceful Uses of Outer Space (UN-COPUOS).

India also plays major role in other multilateral fora including United Nations Economic and Social Commission for Asia and the Pacific (UN-ESCAP), International COSPAS-SARSAT system for search and rescue operations, International Astronautical Federation (IAF), International Academy of Astronautics (IAA), International Institute of Space Law (IISL), Committee on Earth Observation Satellites (CEOS), Committee on Space Research (COSPAR), Inter Agency Debris Coordination Committee (IADC), Space Frequency Coordination Group (SFCG), Coordinating Group on Meteorological Satellites (CGMS),International Space Exploration Coordination Group (ISECG), International Global Observing Strategy (IGOS), International Space University (ISU), Asian Association for Remote Sensing (AARS), International Society for Photogrammetry and Remote Sensing (ISPRS), etc.

In its fourteenth flight conducted from Satish Dhawan Space Centre (SDSC) SHAR, Sriharikota on October 22, 2008, the Indian Space Research Organization's (ISRO's) Polar Satellite Launch Vehicle, PSLV-C11, successfully launched the 1380kg Chandrayaan-1 spacecraft into a transfer orbit with a perigee (nearest point to Earth) of 255 km and an apogee (farthest point to Earth) of 22,860 km, inclined at an angle of 17.9 deg to the equator.

After a 52 hour count down, PSLV-C11 lifted off from the Second Launch Pad at SDSC SHAR at 06:22Hrs Indian Standard Time (IST) with the ignition of the core first stage. The important flight events included the separation of the first stage, ignition of the second stage, separation of the payload fairing at about 116 km altitude after the vehicle had cleared the dense atmosphere, second stage separation, third stage ignition, third stage separation, fourth stage ignition and fourth stage cut-off.[24] Chandrayaan-1 is India's first spacecraft mission beyond Earth's orbit.

It aims to further expand our knowledge about Earth's only natural satellite-the moon. With well-defined objectives, Chandrayaan-1 mission intends to put an unmanned spacecraft into an orbit around the moon and to perform remote sensing of our nearest celestial neighbor for about two years using eleven scientific instruments built in India and five other countries. The primary objectives of Chandrayaan-1 are:

- To place an unmanned spacecraft in an orbit around the moon
- To conduct mineralogical and chemical mapping of the lunar surface
- To upgrade the technological base in the country

Chandrayaan-1 aims to achieve these well-defined objectives through high- resolution remote sensing of moon in the visible, near infrared, microwave and X-ray regions of the electromagnetic spectrum. With this, preparation of a 3-dimensional atlas of the lunar surface and chemical and mineralogical mapping of entire lunar surface is envisaged.

17.3. PROCEDURE OF ESTABLISHING AN ASA

In order to establish an Asian Space Agency we need to take the following five step's procedure.

As a first step, it is necessary to hold a workshop, symposium or Internet mass media assembling space law professors, lawyers, scientists, technicians, high-ranking officials and staff members from the Asian countries' space agencies, including mainly Korea, Japan, China and India in order to concentrate on their opinions concerning the establishment of an ASA.

As a second step, we need to organize a preparatory committee for establishing an ASA through a ministerial conference or diplomatic conference of the Asian countries.

As a third step, a *"Draft Convention for the Establishment of an Asian Space Agency"* should be legislated by excellent law professors or diplomats in collaboration with specialists from the aforementioned Committee.

As a fourth step, after extensive discussion and screening of the Draft by a ministerial or diplomatic conference of all Asian countries, they must pass the Draft by unanimity or by a two-third majority of participants of the ministerial diplomatic conference.

As a fifth step, the Asian countries must ratify "*The Convention for the Establishment of an Asian Space Agency.*"

I would like to propose the following Preamble to the *Draft for Convention*, based on the European Space Agency's Convention[25]

Preamble of the Draft Convention for the Establishment of an Asian Space Agency:

"*Considering* that the magnitude of the human, technical and financial resources required for activities in the space field is such that these resources lie beyond the means of any single Asian country,

Considering that the Resolution will be adopted by the Asian ministerial space conference, which will decide to create a new organization, called the "Asian Space Development Agency" for the development and construction of space vehicle launchers and station, and that the aim would be to integrate the Asian national space programs into an Asian space program as far and as fast as reasonably possible,

Desiring to pursue and to strengthen Asian co-operation, for exclusively peaceful purposes, in space research and technology and their space applications, with a view to their being used for scientific purposes and for operational space application systems,

Desiring in order to achieve these aims, to establish a single Asian space organization to increase the efficiency of the total Asian efforts by making better use of the resources at present devoted to space and to define a Asian space program for exclusively peaceful purposes,(……)."

17.4. PRINCIPAL POINTS THAT NEED TO BE INCLUDED IN THE DRAFT FOR THE CONVENTION

I would like also to propose the following ten principal points that need to be included in the said Draft Convention.

17.4.1. Members and Legal Personality

The members of Asian Space Agency shall be the States parties to the *"Convention for the Establishment of an Asian Space Agency."*

The ASA shall have legal personality. It has the capacity, competence and status of international law, but on the territory of the States Parties to its Convention it also has legal capacity in accordance with national law.

17.4.2. Purpose of Establishment

The purpose of establishing the Asian Space Agency is to provide for and to promote, for exclusively peaceful purposes, cooperation among Asian States in space research and technology and their space application, with a view to their being used for scientific purposes and for operational space applications systems.

The purposes of the ASA shall include in the following particulars;

- Drawing up international rules and monitoring the application of such rules, including the gathering of technical information on space activities conducted under existing legal texts (on registration, recovery liability, satellites with nuclear power resources) or future texts (space shuttle, space station, space debris, etc.……);

- Encouraging the transfer of space technologies to developing countries, the training of specialists, and wide circulation of data gathered in the course of space activities, especially data adapted to the needs of these countries (for ex: distribution of remote sensing data);
- Coordinating environmental monitoring by satellites
- Establishing a monitoring and researching organization in order to protect the environment of earth and space so as to mitigate space debris.

17.4.3. Space Policy

The Asian Space Agency is in charge of elaborating and implementing the medium and long term Asian space policy, of actual activities and programs and a related industrial policy in the space field, and the coordination of Asian and national space programs with respect to international organizations and institutes. Furthermore, the member States decide on Asian assimilation of their national space programs by integrating them into ASA programs. Finally, the Agency elaborates and implements a space industrial policy, which is designed, in particular, to improve the worldwide competitiveness of the Asian space industry.

17.4.4. Exchange of Information

Members and Asian Space Agency shall facilitate the exchange of space policy, programs, scientific and technical information pertaining to the fields of space technology.

17.4.5. Education and Research

The Asian Space Agency shall ensure the execution of basic activities, such as education, documentation, studies of future projects and technological research work. The ASA also facilitates the collection of relevant information and its dissemination to `Member States, assistance and advice for harmonizing national and international programs and the elaboration and execution of scientific programs including the design, development, construction, launching, placing satellites and space shuttle in orbit and control of satellites and all similar activities for launching facilities, space station or space transport system.

17.4.6. International Cooperation

The Asian Space Agency may, upon decisions of the Council taken by vote of a two-thirds majority of all Members States, cooperate with other international organizations and institutions and with Governments, organizations and institutions of non-Member States, and conclude agreements with them to this effect.

17.4.7. Financial Contributions

Asian Space Agency will be financed by its member States. The scale of contributions shall be based on the average national income of each Member State for the three latest years for which statistics are available.

17.4.8. Organs

The Organs of Asian Space Agency are composed General Assembly, Council and the Director General, Senior Staff etc..

17.4.8.1. The General Assembly

1. The participation in the works of the General Assembly is, indeed, open to all States which have accepted the present legal regime.
2. The General Assembly, which is the principal organ, is composed of representatives of all Member States.
3. The meeting of the General Assembly are held every two years. But extraordinary meetings of the General Assembly may be held at any time upon the call of the Council or at the request of any ten contracting States addressed to the Secretary General.
4. It meets when it is required and is composed of either Ministers of the Member's States or government delegates. When it meets at ministerial level it can fulfill the political and practical function for the international conference on the exploitation and development of the moon and other celestial bodies.
5. The General Assembly elects its chairman and its vice-chairman for a period of three years, and re-election is possible for a further year.
6. The Chairman shall direct the meetings, the proceedings, prepare the decisions and maintain appropriate contact with the Member States; he shall advice the Director General and obtain from him all necessary information.
7. When the Council meets at the ministerial level, it shall elect a chairman for that meeting.

17.4.8.2. The Council

1. The Council, which is the principal organ, is composed of representatives from ten
2. Member States. The Council meetings are held annually.
3. It meets when it is required and is composed of either Ministers of the Member's States
4. or government delegates. When it meets at ministerial level it can fulfill the political function of the Asian Space Conference.
5. The Council elects its Chairman and its Vice-Chairman for a period of three year's and re-election is possible for a further year.
6. The Chairman shall direct the meetings, the proceedings, prepare the decisions and maintain and maintain appropriate contact with the Member States; he shall be advice the Director General and obtain from him all necessary information.
7. When the Council meets at the ministerial level, it shall elect a chairman for that Meeting.

17.4.8.3. Director General

- The Director General is the executive of the Asian Space Agency and its representative. The Director General, who is the head of the executive body, is appointed by Chairman of Board of Directors after two-thirds approval resolution is obtained from the Board of Directors.

- He is responsible for the management of the ASA, the execution of the programs
- and he accomplishes all the tasks imposed on him by the Council as well as the implementation of its policy and the attainment of its objectives in accordance with the ASA Convention.

17.4.8.4. Senior Staff

- Members of Senior staff for management, defined by the Council, shall be appointed by Chairman of the Council on the recommendation of the Director General.

17.4.9. Disputes and Arbitration

Disputes between Member States or between any of them and the Asian Space Agency must first be settled by the Council. If the dispute is not settled this way, it shall at request of any party to the dispute be submitted to arbitration.

Unless the Parties agree differently or the Council adopts other rules, the Arbitration Tribunal shall consist of three members. Each Party shall appoint one of them, and those two arbitrators shall designate a third member.

The third member is the umpire and presides over the tribunal. [26)]

The rules of procedure may be agreed between the Parties or imposed by the Council. The award shall be decided by the majority of votes (abstentions are not allowed) and it is final and binding. The Tribunal may also interpret the award at the request of a party to the disputes.

17.4.10. Headquarters

It is desirable that the headquarters of Asian Space Agency are located in the geographical center of the Asian countries, in a city with convenient means of transport. Here meetings of ministers, officials or technical experts from the Member States thrash out big decisions, which the Director General and his senior management give advice about, and then implement.

The directors of ASA's programs for science, applications and launchers are based at the headquarters of ASA, as well as officials responsible for strategy, technological policy, finance etc.

The location of the Asian Space Agency's headquarters is determined by a majority vote of the board of directors among Seoul, Beijing and Tokyo cities.

17.5. CONCLUSION

The Asian Space Agency is regarded as a new road for Asia's space policy and space exploitation in Asian countries. The Asian Space Agency also coordinates the broad thinking needed to meet new challenges in Asian countries. The Asian Space Agency will provide a vision of Asia's future in space, and of the benefits for people on the ground that satellites can supply.

Due to the developments of Internet, telecommunication by the satellites and space stations, it will be extinguished gradually or step by step the boundary among the nations in Asia as well as in Europe including Russia, North and South America, Oceania and Africa during the coming hopeful new Millennium.

I would like to quote a German poem written by the very famous German poet Johann Wolfgang Goethe as the following: *"Wissenschaft und Kunst gehören der Welt an, und vor ihnen verschwinden die Schranken der Nationalität."* [27] I also would like to translate the aforementioned German language poem to English language as the following;

"The Learning and Art belong to the world, and the boundary of nationality is extinguished before them." I believe that the spirit of the said poem is the spirit of space law.

First of all, I am sure that it is possible to establish an electronic Asian Space Agency like electronic Government through Internet as well as an electronic Asian Centre for Space Law as a first step. Since the Asian air and space industry will become a very promising market in the 21st century, we can expect very severe competition among Asian countries and developed countries, such as the USA, Russia, Canada, and EU countries, in order to occupy the Asian market.

To win this severe competition on the Asian air and space industry market, it is necessary for the Asian peoples to work together in union, to strengthen cooperation in research, and to establish friendly relations for the benefit of the air and space industry in all Asian countries. Finally, a very important point is that a political drive, at the highest level, should be given to mobilize states to this initiative, possibly taking the form of a solemn statement by heads of state of Asian countries setting out objectives and prospects for the long term.

It should be noted that this political drive will be necessary not only to set up the organization, but also during a subsequent period.[28] It is desirable and necessary for us to establish the Asian Space Agency, in order to develop the space industry, to strengthen friendly relations and to promote research cooperation among Asian countries based on oriental ideology, ethics and creative ideas. I am sure that it is possible to establish an Asia Space Agency, if the heads of the Asian States would agree to establish the Asian Space Agency through a summit conference.

ADDITIONAL READING

Wüstendörfer. (1950). *Neuzeitliches Seehandelsrecht.* Verlag J.C.B. Mohr. http://yyy.tksc.nasda.go.jp/Home/Press/e/200006/aprsaf_000613_e.html https://vnsc.org.vn/en/news-events/aprsaf-27-will-be-held-in-vietnam-in-2020 http://www.apsco.int/html/comp1/content/NewsHighlights/2019-12-23/54-345-1

KEY TERMS AND DEFINITIONS

APRSAF: The Asia-Pacific Space Cooperation Organization (APSCO) is an inter-governmental organization operated as a non-profit independent body with full international legal status. It is headquartered in Beijing, People's Republic of China. Members include agencies from: Bangladesh, China, Iran, Mongolia, Pakistan, Peru, and Thailand. Indonesia and Turkey also signed the APSCO convention. In 2005, the APSCO convention was signed in Beijing. Representatives from Argentina, Malaysia, the Philippines, Russia and Sri Lanka also attended the founding ceremony.

APSCO: The Asia-Pacific Space Cooperation Organization (APSCO) is an inter-governmental organization operated as a non-profit independent body with full international legal status. It is headquartered in Beijing, People's Republic of China. Members include agencies from: Bangladesh, China,

Iran, Mongolia, Pakistan, Peru, and Thailand. Indonesia and Turkey also signed the APSCO convention. In 2005, the APSCO convention was signed in Beijing. Representatives from Argentina, Malaysia, the Philippines, Russia and Sri Lanka also attended the founding ceremony.

ISRO: The Indian Space Research Organization (ISRO) is the national space agency of the Republic of India, headquartered in Bengaluru. It operates under Department of Space (DoS) which is directly overseen by the prime minister of India while chairman of ISRO acts as executive of DoS as well. ISRO is the primary agency in India to perform tasks related to space-based applications, space exploration and development of related technologies. It is one of six government space agencies in the world which possess full launch capabilities, deploy cryogenic engines, launch extraterrestrial missions and operate large fleets of artificial satellites. ISRO launched its second lunar mission Chandrayaan-2 to study the lunar geology and the distribution of lunar water. ISRO's Mission to Venus is scheduled for 2025 that will include a payload instrument called Venus Infrared Atmospheric Gases Linker (VIRAL) which is co-developed with Laboratoire atmosphères, milieux, observations spatiales (LATMOS) under French National Centre for Scientific Research (CNRS) and Roscosmos. The next Mars mission, Mars Orbiter Mission 2, also called *Mangalyaan 2* (Sanskrit), will be launched in 2024. It will have a less elliptical orbit around Mars and could weigh seven times more than the first mission. This orbiter mission will facilitate the community to address several open science problems. The science payload of the planned satellite is estimated at no more than 100 kg (220 lb.).

ENDNOTES

1 http://www.nasa.gov

2 http://liftoff.msfc.nasa.gov/rsa/rsa.html

3 http://www.esa.int/export/esaCP/index.html

4 APE Cooperation (APEC) was established in 1989 in response to the growing interdependence among Asia-Pacific economies. The APEC has since become the primary regional vehicle for promoting open trade and practical economic cooperation. Brunei Darussalam is the APEC Chair for the year 2000. Members: Australia, Brunei Darussalam, Canada, Chile, People's Republic of China, Hong Kong (China), Indonesia, Japan, Korea, Malaysia, Mexico, New Zealand, Papua New Guinea, Peru, Philippines, Russia, Singapore, Chinese Taipei, Thailand, United States of America, Viet Nam.

5 The Association of South East Asian Nations (ASEAN) was established August 8, 1967 in Bangkok, Thailand by the five original Member Countries, namely Indonesia, Malaysia, Philippines, Singapore, and Thailand. Brunei Darussalam joined on January 8, 1984, Viet Nam on July 28, 1995, Laos and Myanmar on July 23, 1997 and Cambodia on April 30, 1999. The ASEAN region has a population of about 500 million, a total area of 4.5 million square kilometers, a combined gross domestic product of US$ 737 billion, and a total trade of US $720 billion. In 1995, the ASEAN Heads of State and Government re-affirmed that cooperative peace and shared prosperity shall be the fundamental goals of ASEAN.

6 ASEM. the acronym for Asia-Europe Meeting, is a biennial Summit meeting of heads of state and government of 10 Asian countries, 16 EU member states, including the president of the European Commission. In order to facilitate cooperation in various fields, in addition to the biennial Summit meetings, other meetings such as Foreign Ministers' Meeting, Economy & Finance Ministers

'Meeting and Senior Officials' Meeting are held. The first Asia-Europe Meeting (ASEM I) was inaugurated in Bangkok on March 1996. ASEM II was held in London on April 1998. ASEM III was held successfully in Seoul in October 2000. ASEM IV will be held in Copenhagen, in 2002. Members: Asia (10): Brunei, China, Indonesia, Japan, Malaysia, Philippines, Republic of Korea, Singapore, Thailand, Viet Nam. Members: Europe (16): Austria, Belgium, Denmark, Finland, France, Germany, Greece, Ireland, Italy, Luxembourg, Netherlands, Portugal, Spain, Sweden, United Kingdom, EU Commission.

7 The Korean government announced recently that by 2005 it will construct a Space Center for the launching of satellites at Oenaro Island located near the south coast of the Korean Peninsula. The "Space Center" will serve as the infrastructure for space and technological development and plan to launch a low earth orbit satellite in 2008. A second science satellite made in Korea will be launched from the space center by 2005 and by 2015 four multi-purpose and five science satellites will be launched. It was added that from 2010 the center will be operated on a commercial basis operating launch facilities for low to mid altitude orbit satellites.

8 http://yyy.tksc.nasda.go.jp/Home/Press/e/200006/aprsaf_000613_e.html

9 https://www.aprsaf.org/annual_meetings/aprsaf27/meeting_details.php

10 https://www.jaxa.jp/about/int/index_e.html

11 http://www.cnsa.gov.cn/administor_message.htm

12 http://www.costind.gov.cn/e_index.htm

13 https://www.unescap.org

14 http://www.cnsa.gov.cn/policy_space.htm

15 https://en.wikipedia.org/wiki/Shenzhou_7

16 https://en.wikipedia.org/wiki/Shenzhou_10

17 https://en.wikipedia.org/wiki/Shenzhou_11

18 https://en.wikipedia.org/wiki/Shenzhou_12

19 https://en.wikipedia.org/wiki/Asia-Pacific_Space_Cooperation_Organization

20 http://www.cnsa.gov.cn/n615709/n620682/n639462/79381.html

21 http://www.cmmacs.ernet.in/nal/icast/isro.html

22 http://www.isro.org

23 https://www.isro.gov.in/international-cooperation

24 http://www.isro.org/pressrelease/Oct22_2008.htm

25 In 1975, European Space Conference, meeting in Brussels, approved the text of the Convention for the Establishment of a European Space Agency" setting up the European Space Agency. The member states are now fifteen countries: Austria, Belgium, Denmark, Finland, France, Germany, Ireland, Italy, the Netherlands, Norway, Portugal, Spain, Sweden, Switzerland, and the United Kingdom. Canada is a Cooperating State: United Nations, "Space Activities of the United Nations and International Organizations", UN (New York, 1992), at 135; H.L. van Traa-Engelman, "Commercial Utilization of Outer Space," Martinus Nijhoff Publishers (1993), at 160-161.

26 E. R. C. Van Bogaert, "Aspects of Space Law", Kluwer, (1986), at 271.

27 Dr. jur. Hans Wüstendörfer, "Neuzeitliches Seehandelsrecht", Verlag J.C.B. Mohr (Paul Siebeck), Tübingen, (1950), S. 17.

28 Gabriel Lafferranderie, "Outlook on Space Law over the next 30 years", Essays Published for the 30th Anniversary of the Outer Space Treaty, Kluwer Law International (1997), at 427.

Chapter 18
Proposal of Establishing a New International Space Agency for Mining the Natural Resources on the Moon, Mars, and Other Celestial Bodies

ABSTRACT

This chapter explains the author's proposal on the necessity of "Establishing a New International Space Agency for Mining the Natural Resources in the Moon, Mars, Asteroid, and Other Celestial Bodies." The International Space Agency (ISA) will be regarded as a new road for the global space policy and exploitation of the moon and other celestial bodies in the global community. As the moon, Mars, asteroid, Saturn, Jupiter, Titan, and other celestial bodies have a large quantity of the precious natural resources, we must establish a new ISA in order to explore and develop efficiently and effectively the aforementioned minerals. It is necessary for us to establish the ISA so as to work together in union to strengthen cooperation in research and to establish friendly relations for the benefit of mankind. Finally, a very important point is that a political drive, at the highest level, should be given to mobilize states to this initiative, possibly taking the form of a solemn statement by heads of the space superpower countries setting out objectives and prospects for the long term.

18.1. INTRODUCTION

The Moon Agreement provide that the moon and its natural resources are the common heritage of mankind and the harvesting of those resources is necessary for the new international regime established to govern the exploitation of such resources when it becomes feasible to do so. The establishment of a new International Space Agency (ISA: tentative name) would lead to a strengthening of the cooperation deemed essential by the global community towards joint undertakings and development of the natural

DOI: 10.4018/978-1-7998-7407-2.ch018

resources of the moon, Mars, asteroid, Saturn, Jupiter, Titan and Other Celestial Bodies. The new ISA would act as a catalyst for the efforts on space development and allow natural resources, technology, manpower and finances to be centrally managed in an independent fashion to the benefit of the mankind.

"Space already exists for the global community, so the question is what we do there."

It's our job to make sure that all opportunities are used to integrate the power and exploitation of the moon, Mars, asteroid, Venus, Jupiter and Other Celestial Bodies among the global all countries. In the 21st century, space science and technology will develop with ever greater rapidity.

Having developed rapidly for about half a century, some space superpowers and organization such as the United States, EU, Russia, China, Japan, India with space activities have scored remarkable achievements that greatly promoted the development of social productivity and progress.

The continuous development and application of space technology has become an important role in the modernization drive of the world community.

The emergence of aerospace technology in the global countries has brought huge contributions to economic and social progress.

Nevertheless, in order to allow such an development to start a major obstacle must be solved, namely the absence within the space law regime of specific rules establishing how this exploitation has to take place and what are the rights and duties of the parties involved in it. As it will be explained in the following paragraphs, space law does not contain any dedicated rule, dealing with the exploitation of extraterrestrial resources such as the moon, Mars, asteroid, Venus Jupiter, which has received the general acceptance of States.[1]

Since the 1967 Space Treaty and the 1979 Moon Agreement did not provided any provisions on the following three items such as procedure of getting mining license, essence of mining right, right and duty of miners, so the United Nations Committee on the Peaceful Use of Outer Space (UNCOPUOUS) must revise the 1967 Space Treaty and the 1979 Moon Agreement so as to include the following three items in the near future.

1. Procedures and registration for the permits of mining right on the natural resources in the moon, Mars, asteroids, Jupiter, Saturn and Other Celestial Bodies
2. Essence of mining right, the duration of the mining license, the rights and obligations of the mining license holder,
3. Cooperation among the ISA member states each other for mining natural resources on the moon, Mars, asteroids, Saturn and Other Celestial Bodies.

18.2. DEVELOPMENT OF THE NATURAL RESOURCES IN THE MOON, MARS AND OTHER CELESTIAL BODIES

18.2.1. Development of the Natural Resources Helium-3 on the Moon

The moon, Mars[2], asteroid[3], Saturn[4], Jupiter,[5] Titan[6] and Other Celestial Bodies of our solar system contained large quantity of natural resources. The development of the natural resources on the moon and Other Celestial Bodies represents one of the most existing future developments in the field of space law as well as a unique occasion for the economic and social growth of mankind as a whole. As it is well-known, mankind is currently facing an energetic crisis.

The large number of benefits that are expected to be generated from the development of these resources, indeed, not only will contribute to the betterment of conditions of peoples on earth but also will allow mankind to face and likely solve one of the biggest problems currently affecting our planet, namely the exhaustion of the stocks of raw materials and other source of energy, such as fossil fuels.

There is no doubt that one of the most difficult problems that a peaceful world will face in the 21st century will be to secure an adequate, safe, clean, and economical source of energy. Existence of lunar Helium-3, to be used as fuel for fusion reactors, is well documented; verified from numerous Apollo and Luna mission samples, current analyses indicate that there are at least 1 million tones embedded in the lunar surface. The Helium-3 would be used as fuel for fusion reactors. Moon gas may solve earth's energy crisis.

If the current trends of energy on earth, scientists anticipate that the energy resources oil after 50 years, natural gas after 60 years, and uranium after 65 years will be dried up. So the research of the nuclear fusion for electric generator is progressed for long time. The solar energy is similar to creating the *'artificial sun'*.

A scientist warns of the exhaustion of fossil fuels such as coal, oil and natural gas on earth. By 2060 the whole world will have a major problem. We need to be thinking ahead.

Right now we are not thinking ahead enough. Scientists estimate there is about 1 million tons of Helium-3 on the moon, enough to power the world for thousands of years. The equivalent of a single space shuttle load or roughly 25 tons could supply the entire United States' energy needs for a year, according to Apollo17 astronaut and researcher Mr. Harrison Schmitt of the Fusion Technology Institute (FTI).[7]

It has been estimated that 25 tons of Helium-3 can provide all the power that the United States needs in a year. The stocks of raw materials are running out and experts estimate that fossil fuels will be finished by 40~50 years. Helium-3, indeed, has the potential to solve this crisis thanks to its capacity to replace fossil fuels and other substances as primary source of energy on earth.

As to the moon, it presents vast amount of mineral resources distributed uniformly across its surface and subsurface. Manned and unmanned explorations have demonstrated that the moon is rich of ① aluminum, ② iron, ③ silicon, ④ oxygen, ⑤ hydrogen, ⑥ chromium, ⑦ manganese, ⑧ potassium, etc.

These minerals can be utilized in their original form or refined into structural and electrical materials. They can be brought back to earth or used for life support of a permanent lunar basis or as rocket propellant.[8] For instance, oxygen and hydrogen are contained in the lunar regolith at all latitudes. There is also evidence that the lunar poles contain amounts of water and ice.

Lunar water is water that is present on the Moon. Liquid water cannot persist at the Moon's surface, and water vapor is decomposed by sunlight, with hydrogen quickly lost to outer space. However, scientists have since the 1960s conjectured that water ice could survive in cold, permanently shadowed craters at the Moon's poles. Water molecules are also detected in the thin layer of gases above the lunar surface.[9]

It is still not well-known how vast this amount is. However, in case of a large presence of water, this could have an enormous impact as rocket propellant and life-support materials for astronauts. A potential gas source found on the moon's surface could hold the key to meeting future energy demands as the earth's fossil fuels dry up in the coming decades.

When compared to the earth the moon has a tremendous amount of Helium-3," when Helium-3 combines with deuterium (an isotope of hydrogen) the fusion reaction proceeds at a very high temperature and it can produce awesome amounts of energy.

The most valuable resource contained on the moon, however, is Helium-3.

The Helium-3 represents, indeed, the main reason behind the attention of States and private operators for exploiting extraterrestrial resources.

The raw material Heliumn-3 for the Nuclear Fusion Reactor is not embedded in the earth, but it is estimated to be embedded 1 million tonnes~500 million tons in the lunar surface. Approximately one million tons of Helium-3 is a quantity to use and create the energy for 500 years in the global community.

Helium-3 is an isotope, rare on earth but abundant on the moon, which combined with other materi-als, such as deuterium, can be used as a fuel in fusion power reactors.

The value of Helium-3 is that it can generate nuclear power and, as a consequence, energy in a clean way, namely through a process of nuclear fusion which does not produce toxic waste.

Thanks to these special characteristics, the extraction of Helium-3 is likely to have a huge impact on the way energy is produced and distributed on earth. Helium-3 is deposited on the lunar surface by solar winds and would have to be extracted from moon soil and rocks.

To extract Helium-3 gas the rocks have to be heated above 800 degrees Celsius. The 200 million tons of lunar soil would produce one tone of Helium. Only 10 kilograms of Helium-3 are available on earth.[10] As space superpowers such as the United States, Russia, European Union, China, Japan and India has interested in Helium-3 that it is more expensive 300 times than gold, so it may be become a supremacy country in the future resources war from mining Helium-3 in the moon to bring it to earth for the purpose of getting it in advance. The Helium-3 fusion energy may be the key to future space exploration and settlement.

As we must promote the development of the moon's natural resources including Helium-3 in order to resolve in advance the depletion of energy resources in the global community after 50-60 years, so it is absolutely necessary the active and reciprocal cooperation of space science and technology in the first place among the space superpowers.

Figure 1.
Source: https://missiontothegrey.weebly.com/helium-3.html

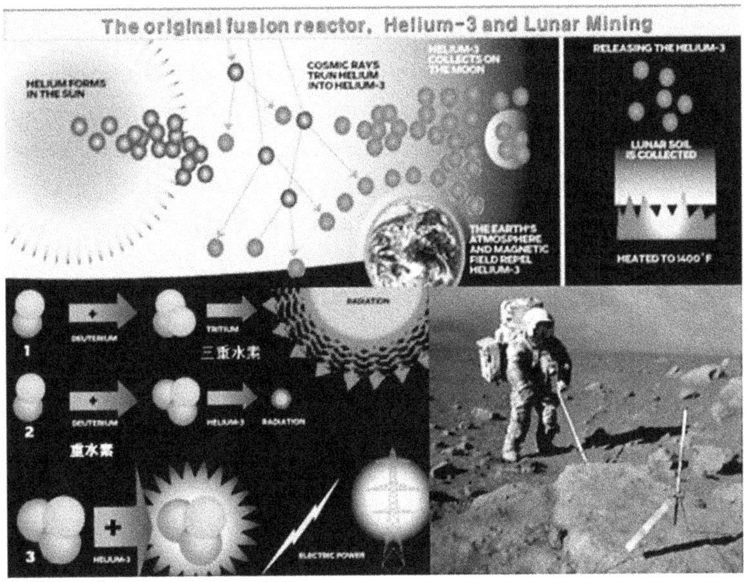

12.2.2. The Moon Has a Lot of Ice

NASA of the United States has confirmed the ice on the moon for the first time and hopes that manned exploration will resume. Administrator Jim Bridenstine, Secretary of the National Aeronautics and Space Administration (NASA), said on August 21, 2018, that the presence of ice on the surface of the moon confirmed his vision of resuming lunar exploration in a *"sustainable"* form. Using data from a NASA radar that flew aboard India's Chandrayaan-1 spacecraft, scientists have detected ice deposits near the moon's north pole.

NASA's Mini-SAR instrument, a lightweight, synthetic aperture radar, found more than 40 small craters with water ice. The craters range in size from 1 to 9 miles (2 to15 km) in diameter. Although the total amount of ice depends on its thickness in each crater, it's estimated there could be at least 1.3 trillion pounds (600 million metric tons) of water ice.[11] The Mini-SAR has imaged many of the permanently shadowed regions that exist at both poles of the Moons. These dark areas are extremely cold and it has been hypothesized that volatile material, including water ice, could be present in quantity here.

Figure 2. NASA: Ice Confirmed Moon's Pole
Source: https://www.space.com/7987-tons-water-ice-moon-north-pole.html

NASA Administrator Jim Bridenstine announced October 25, 1999 that the agency would send a robotic rover to the moon in 2022 to look for water ice, confirming plans that had been taking shape for months. In a speech at the 70th International Astronautical Congress, Bridenstine said the Volatiles Investigating Polar Exploration Rover (VIPER) mission would look for ice on or below the surface of the moon at its south pole, a key resource for future human missions.[12]

18.2.3. Huge Ice on the Mars' Natural Resources and Crater, ESA Processes Data and Releases Images

Mars looks red because there is no sea of water like earth on its surface, and the earth's surface contains a large amount of iron oxide (red rust). The composition of Mars' atmosphere is 95% carbon dioxide, 3% nitrogen, 1.6% argon, and contains minor components such as oxygen and water vapor. Mars also has a methane gas source. The surface of Mars is mainly composed of basalt and andesite rocks. Both are rocks on the earth where magma forms by solidifying near the surface of the earth and are distinguished by the amount of silicon dioxide (SiO2) contained.

Many places on Mars are covered with fine dust such as talc powder a few meters or more in thickness. Small spherical hematite (hematite) was discovered in rocks collected on the Meridian Plain on Mars. The sphere is only a few millimeters in diameter and is thought to have been formed billions of years ago as a sedimentary rock in a wet environment. Other minerals containing sulfur, iron, and bromine have been discovered in the Mars.[13]An image of Mars' crater with lots of ice and snow left was captured and released. A large amount of snow and water ice were captured on the "Korolev crater" near the North Pole of Mars.

This image was taken by the Mars Express, a Mars spacecraft from the European Space Agency (ESA). [14]It has been in orbit over Mars since 2003, and the image was taken in April 2018 by its onboard high-resolution stereo camera (HRSC). Pathfinder's Sojourner Rover measured elements in Mars rocks with NASA's Alpha Proton X-ray spectrometer. Many spacecraft explorations of Mars have shown that the red planet has natural resources essential to sustain human life.

Important life support compounds CO2, Ar, N2 and H2O are available on Mars. Solar energy, and possibly wind power, is practical as it is available on Mars. Mars may contain ores that would be very useful to potential colonist.[15] The abundance of volcanic features together with widespread cratering are strong evidence for a variety of ores.

While nothing may be found on Mars that would justify the high cost of transport to Earth, the more necessary ores future colonists can obtain from Mars, the easier it would be to build colonies on the Red Planet.[16] Ore deposits are produced with the help of large amounts of heat. On Mars, heat can come from molten rock moving under the ground and from crater impacts. Liquid rock under the ground is called magma. When magma sits in underground chambers, slowly cooling over thousands of years, heavier elements sink. These elements, including copper, chromium, iron, and nickel become concentrated at the bottom.[17] When magma is hot, many elements are free to move. As cooling proceeds, the elements bind with each other to form chemical compounds or minerals.

Because some elements do not bond easily to form minerals, they exist freely after nearly all the other elements have bonded into compounds or minerals. The remaining elements are called incompatible elements. Some of them are quite useful to humans.

Some examples include niobium a metal used in producing superconductors specialty steels, lanthanum and neodymium and europium for television monitors and energy-efficient LED light bulbs.[18]

After the mass of magma has cooled and has mostly frozen or crystallized into a solid, a small amount of liquid rock remains. This liquid bears important substances such as lead, silver, tin, bismuth and antimony.[19] Sometimes minerals in the magma chamber are so hot that they occupy a gaseous state. Others are mixed with water and sulfur in aqueous solutions. The gases and mineral-rich solutions eventually work their way into cracks and become useful mineral.

Ore minerals, including the incompatible elements, remain dissolved in the hot solution, then crystallize out when the solution cools. Deposits created by means of these hot solutions are called hydrothermal deposits. Some of the world's most significant deposits of gold, silver, lead, mercury, zinc, and tungsten started out this way.[20]

It is desirable for us the establishment of the ISA in order to be efficient and rapid development among the abovementioned countries so as to manage, allotment and the adjustment for the development and exploration of moon natural resources including Helium-3.

The creation of ISA is possible to promote the unification of the window for negotiations in cooperating of the space science and technology among the United Nations Committee on the Peaceful Use of Outer Space (UNCOPUOS), the United States (National Aeronautics and Space Administration: NASA), European Union (European Space Agency: ESA), China (China National Space Administration: CNSA), Japan (Japan Aerospace Exploration Agency: JAXA) and India (Indian Space Research Organization; ISRO) including Korea (Korea Aerospace Research Institute: KARI) etc. for exploiting of the moon and other celestial bodies' natural resources.

Figure 3. Mars Crater
Source: https://www.universetoday.com/140958/this-crater-on-mars-traps-the-cold-and-remains-filled-with-ice-all-year-round

Moreover it is necessary and fruitful for mankind to establish the new ISA as soon as possible. If it is established the ISA in the near future, the ISA will be developed many the natural resources of the moon in cooperation with the developed countries and is able to create a fund to collect 10 percent commission for the authorization of mining right them in the moon due to the mankind's common heritage for the moon's natural resources based on the Article 11 of the 1979 Moon Agreement.

The ISA will be used these funds for the economic aid and support for space exploration and research in developing countries.

18.2.4. Asteroids and Natural Resources

Asteroids can be classified as consisting of carbon, silicon, metal, gold or diamond by astronomical telescope observations were taken along with an analysis of meteorites (asteroid fragments that fell on Earth). And astronomical data suggests that a small percentage of asteroids contain high concentrations of valuable metals, such as platinum and gold.

First of all, we need to find the right asteroid mining target. To date, more than 750,000 asteroids have been identified. hey are found in major asteroid belts and orbit between Mars and Jupiter, but that's too much and potential mining targets should be considered. We should focus on asteroids near earth. A trillion dollars worth of resources may be asleep on an asteroid. According to the astrophysicist Professor Martin Elvis (Harvard University)[21], a mining -worthy asteroid has an market value $ 1billion[22]

Figure 4. Asteroids Belt
Source: https://spaceplace.nasa.gov/asteroid-or-meteor/en

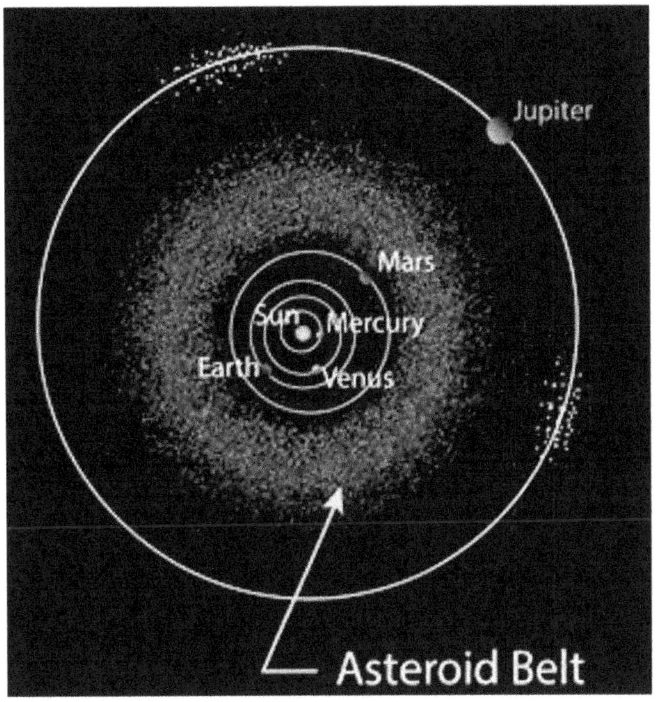

18.2.5. Saturn and Natural Resources

Saturn is the sixth planet from the Sun and the second largest planet in our solar system. Adorned with thousands of beautiful ringlets, Saturn is unique among the planets. It is not the only planet to have rings—made of chunks of ice and rock—but none are as spectacular or as complicated as Saturn's. Like fellow gas giant Jupiter, Saturn is a massive ball made mostly of hydrogen and helium.[23]

The little dark spot on Saturn is the shadow from Saturn's moon Enceladus.[24] Saturn has a central nucleus made of rock, a compound of iron, nickel and silicon and oxygen, which is thought to be covered by metallic hydrogen thickly. The middle layer contains liquid hydrogen and helium. The outside is surrounded by gas.[25]

According to observations by NASA's Saturn observation satellite "Casini", Saturn's satellite "Titan"" ' has liquefied hydrocarbons that exceed the reserves of the earth's petroleum resources. According to a paper published in the specialist journal "Geophysical Research Letters", it became clear that Hydrocarbons are sometimes the main components of oil and natural gas, and the Titan satellite is a treasure trove of energy resources that surpass the earth.

The scientist who conducted the study was Dr. Ralph Lorenz of the Institute of Applied Physics at John Hopkins University in the United States.

According to Dr. Lorentz, liquefied hydrocarbons are methane on the surface of the satellite "Titan". It is said to have dozens of lakes as ethane and ethane, and that this one lake alone is equivalent to about 130 billion tons of natural gas reserves on Earth.[26]

18.2.6. Joint Exploitation of Natural Resources in the Moon and Legal Basis on the Establishment of ISA

In the near future, if the ISA is established in accordance with Article 11, paragraphs 5 to 6 of the 1979 Moon Agreement, the ISA will be able to develop and exploit many natural resources such as aluminum, iron, silicon, oxygen, hydrogen, chromium, manganese and Helium-3 etc. in the moon and other celestial bodies.

Article 11, paragraphs 1 to 2 of the Moon Agreement stipulates as follows.

1. The moon and its natural resources are the common heritage of mankid,
2. The moon is not subject to national appropriation by any claim of sovereignty, by means of use or occupation, or by any other means.

18.2.7. Cooperative Exploitation of Natural Resources in the Moon and Establishment of ISA

As the ISA will work with developed countries to develope a lot of natural resources in the moon and Other Celstial odies, so when granting mining rights, it is necessary to collect 10-20% commission and accmulate this amount as a fund. The ISA can use the fund to provide financial assistance and space exploration and research to developing countries including African countries.

The establishment of the International Space Agency (ISA) requires discussions to establish a space for science and technology cooperation between the United Nations Commission on Space and Peace (UNCOPUOS) and the United States (National Aeronautics and Space Administration: NASA). In Europe, the European Space Agency (ESA), China National Space Administration (CNSA), Japan Aerospace Exploration Agency (JAXA), Indian Space Research Organization (JAXA) ISRO) and Korea Aerospace Research Institute (KARI) should also discuss the matter in preparation for the establishment of the ISA.

The establishment of an ISA is essential for efficient and rapid development among the above countries to manage, allocate and coordinate the development and exploration of lunar natural resources, including Helium-3.

18.2.8. International Trends on the Recent Space Mining

On March 18, 2016, the European Center for Space Law (ECSL) held a forum in Paris under the title "*Development of Natural Resources in Space from a Scientific, Technical and Legal Perspective*".

The issue of space resource attribution has again sparked heated debate. The 56th Legal Subcommittee of the UN Committee on the Peaceful Use of Outer Space (UNCOPUOS), was held in Vienna, Austria from March 27 to April 7, 2017.[27]

During the first official meeting on the topic *"General exchange of ideas on potential legal models for operations in the exploration, development and use of peace and use of space resources*", the International Space Law Institute (IISL) and Europe A space law symposium jointly hosted by the Space Center was held.

The theme was "*View of the legal model for exploring the use and utilization of space resources for 50 years after the adoption of the Space Treaty.*" The 57th session of the Legal Subcommittee of the United Nations Committee on Peaceful Uses of Outer Space (UNCOPUOS) was taken place April 9-20,

2018 in in Vienna, Austria. At the 57th Legal Subcommittee of the UNCOPUOS, a Dutch delegation introduced the progress of the Hague Working Group of 18 countries.

During the session, the Hague Working Group held a side meeting with the University of Vienna to discuss topics from the Legal Subcommittee of the UNCOPUOS on the exploration, development, and use of space resources.

They exchanged opinions on the legal model and introduced the "*Draft Module for the Development of an International Framework for Space Resources Activities*".

The Legal Subcommittee of the UNCOPUOS was held in Vienna, Austria from 1 April to 12 April, 2019.

The discussions on the legal norms related to the exploration, development and use of space resources require the involvement of not only governments but also the private sector. It is necessary for us to promote the development and exploiutationn of space resources related industries. We welcomed discussions in The Hague Space Resource Governance Working Group. Japan Aerospace Exploration Agency (JAXA) will join the Working Group as an observer from 2019.[28]

18.3. LEGAL PROBLEMS AND SOLUTION FOR DEVELOPMENT OF THE NATURAL RESOURCES AND OTHER CELESTIAL BODIES

Recently it is most severe competition among the space superpowers in order to mine and exploit the natural resources including Helium-3 from the moon so as to solve the serious problems of the earth's energy. As it is un-ratified by any major space-faring powers and unsigned by most of them, it is of no direct relevance to current space activities. The space superpowers and private operators have not started to exploit the resources of the Moon and Other Celestial Bodies yet is the absence of rules setting out how this exploitation shall be carried out. The space law system, indeed, does not provide any specific rule, relating to the exploitation of extraterrestrial resources, which have been generally accepted by States.

On one side, the Outer Space Treaty does not contain any mention of space natural resources or to their possible exploitation. On the other side, the Moon Agreement, whose main purpose is to set forth rules aimed at regulating the use for scientific and commercial reasons of lunar and other celestial bodies' materials, has been rejected by the majority. According to the 1967 Outer Space Treaty (OST)[29] and the 1979 Moon Agreement, [30]

Agreement, these two instruments does not offer an adequate legal framework which is able to ensure the safe, orderly and peaceful development and extraction from the natural resources of the Moon and other celestial bodies. of States, comprising the space-faring States.[31]As a consequence, its principles lose relevance when applied to the exploitation of extraterrestrial resources.

The major problem of the Space Treaty and Moon Agreement is that it does not contain any specific reference to space resources and to their exploitation. But Prof. Dr. Ram Jakhu often recommended "All States should ratify the Moon Agreement as soon as possible."

The exploitation of moon materials raises several specific legal issues, such as those related to the right of mining extraterrestrial sites or to property rights over the extracted materials, which may not be properly dealt with and solved by simply relying on the existing space law principles.

It is realistic to anticipate that the exploitation of these mineral resources will take place in a three phase process: ① pre mining phase; ② mining phase; ③ post mining phase.[32]The rules regulating the exploitation of the resources of the Moon and other celestial bodies should be inserted in a legal instrument which will be opened for acceptance by State and International Organizations.

The ISA is the international organization through which States manage and control the exploitation of the natural resources of the Moon and Other Celestial Bodies.

The ISA has the power to authorize persons to exploit for commercial purposes a certain lunar or other celestial bodies' area. At the same time, however, the ISA has the duty to control that the exploitative activities are carried out in accordance with the space law principles and in a not detrimental manner for the space environment.

18.4. PROCEDURE FOR ESTABLISHING OF THE NEW INTERNATIONAL SPACE AGENCY

The ISA is the international organization through which States manage and control the exploitation of the natural resources of the Moon and Other Celestial Bodies.

The establishment of ISA as a new international regime is based on the Article 11, 5[33] and Article 18[34] of the 1979 Moon Agreement. In order to establish as a preliminary procedure, it needs to make the Draft for the Convention on the establishment an ISA among the space-faring countries. This provision shall be implemented in accordance with article 18 of this Agreement. It is necessary for us to take the following five step's procedure in order to create an ISA.

As a first step, it is necessary to hold a workshop, symposium or Internet mass media assembling space law professors, lawyers, scientists, technicians, high-ranking officials and staff members from the global space agencies such as the UNCOPUOS, the United States (NASA), European Union (ESA), China (CNSA), Japan (JAXA), India (ISRO) and South Korea (KARI) etc. in order to concentrate on their opinions concerning the establishment of the ISA.

As a second step, we need to organize a preparatory committee for establishing the ISA through a ministerial conference or diplomatic conference of the space powers countries including delegate of the United Nations Committee on the Peaceful Use of Outer Space (UNCOPUOS).

As a third step, a *"Draft for the Convention on the Establishment of an International Space Agency"* should be legislated by excellent space law professors, lawyers, space scientists or diplomats in collaboration with specialists from the aforementioned Committee.

As a fourth step, after extensive discussion and screening of the abovementioned *"Draft for the Convention for the Establishment of an International Space Agency."* by diplomatic conference in the UNCOPUOS, they must pass the abovementioned *"Draft for the Convention"* by two-third majority of diplomatic conference in the UNCOPUOS and UN.

As a fifth step, the UN member's countries would like to ratify *"The Convention for the Establishment of an International Space Agency."* I would like to propose the following Preamble to the *"Draft for the* Convention," base on the International Space Agency.[35]

Preamble of the Draft for the Convention on the Establishment of an International Space Agency

The States Parties to this Agreement,

Recognizing the common interest of all mankind in the progress of the exploration and use of outer space for peaceful purposes,

Determined to promote on the basis of equality the further development of co-operation among States in the exploration and use of the moon and other celestial bodies,

Desiring to contribute to broad international co-operation in the scientific as well as the legal aspects of the exploration and use of the moon and other celestial bodies for peaceful purposes,

Desiring to prevent the moon and other celestial bodies from becoming an area of international conflict and environmental damage,

Bearing in mind the benefits which may be derived from the exploitation of the natural resources of the moon and other celestial bodies,

Recalling the contents and essence of Treaty on Principles Governing the Activities of States in the Exploration and Use of Outer Space, including the Moon and Other Celestial Bodies (1967), the Agreement on the Rescue of Astronauts, the Return of Astronauts and the Return of Objects Launched into Outer Space (1968), the Convention on International Liability for Damage Caused by Space Objects (1972), the Convention on Registration of Objects Launched into Outer Space (1975) and Agreement Governing the Activities of States on the Moon and Other Celestial Bodies (1979).

Taking into account the need to define and develop the provisions of these international instruments in relation to the moon and other celestial bodies, having regard to further progress in the exploration and use of the moon and outer space.

18.5. THE MAIN ITEMS THAT NEED TO BE INCLUDED IN THE DRAFT FOR THE CONVENTION ON THE ESTABLISHMENT OF ISA

I would like also to propose the following ten principal points that need to be included in the said "*Draft for the Convention.*

18.5.1. Members and Legal Personality

The members of ISA shall be the States parties to the *"Convention for the Establishment of an International Space Agency."* The ISA shall have legal personality. The ISA may exercise also its functions and powers, as provided in this Statute, on the territory of any State Party and, by special agreement, on the territory of any other State. It shall also have such legal capacity as may be necessary for the exercise of its functions and the fulfillment of its purposes.

18.5.2. Purpose of Establishment

The purpose of establishing the ISA is to provide for and to promote, for exclusively peaceful purposes, cooperation among the global States in space research and technology and their space application for the moon, Mars and other celestial bodies, with a view to their being used for scientific purposes and for operational space applications systems.

The purposes of the ISA shall include in particular:

- *Drawing up* international rules and monitoring the application of such rules, including the gathering of technical information on space activities conducted under existing legal texts (on license, registration, recovery liability, satellites with nuclear power resources) or future texts (space shuttle, space station, space debris, etc.......);
- *Encouraging the transfer* of space technologies to developing countries, the training of specialists, and wide circulation of data gathered in the course of space activities, especially data adapted to the needs of these countries (for ex: distribution of remote sensing data);
- *Coordinating* environmental monitoring by satellites and spacecraft
- Establishing a monitoring and researching organization in order to protect the environment of earth and space so as to mitigate space debris.

18.5.3. Space Policy

The ISA is in charge of elaborating and implementing the medium and long-term global space policy, of actual activities and programs and a related industrial policy in the space field, and the coordination of UN member's states and their national space programs with respect to international organizations and institutes.

Furthermore, the member States decide on global assimilation of their national space programs by integrating them into the ISA programs. Finally, the ISA elaborates and implements a space industrial policy, which is designed, in particular, to improve the exploitation and developments of the global space industry for the moon and other celestial bodies.

18.5.4. Licensing of Mining on the Natural Resources

Unexploited minerals in the moon and other celestial bodies shall not be mined without establishment of mining rights. The license of mining the large quantity of the natural resources in the lunar or other celestial bodies will be bestowed by the ISA.

Persons wishing to secure establishment of mining rights shall apply to and obtain approval of the ISA.

The applicant of receiving the mining license must submit application document with document including

1. Applicant's name, birth day, address and occupation, plan (time schedule) of the exploitation minerals on the moon and other celestial bodies by applicant, indication for the geographical location of the extraterrestrial site as an object of the license,
2. compliance with regulation of ISA and international space conventions by applicant to ISA. The ISA, by means of the Council, has the duty to control the operator of the licensee[36].

In case such a control shows that licensee has not respected the terms of the license, the Council could ask the licensee to stop these violations and to take the required measures.

18.5.5. Mineral Right

The term, "mineral right" means the right to mine or acquire registered minerals, Helium-3 and other minerals buried in the same mineral deposits as the registered minerals, within a rock regolith district of the moon and other celestial bodies (hereinafter referred to as "mining area"). Minerals separated from the mining area without mining rights or mining concession rights shall be owned by the mining right holders or mining lease right holders.

The legal nature of mineral rights is the property rights. If a mining applicant receives notice of approval of establishment of a mineral right, he shall pay registration fee under the conditions as prescribed by the ISA regulation, and apply for registration to the General Secretary of ISA within 30 days from the date of receiving the notice of approval. A mineral right holder or any other interested person may apply to the ISA a survey of the boundaries of his mining area or the adjacent mining area.

18.5.6. Duration of Mineral Rights

The duration of a mineral right shall not exceed 25 years. A mineral right holder may extend the term of a mineral right with the approval of the ISA before the expiration of such term, under the conditions as prescribed by the ISA regulations. In this case, such renewal shall be made for a period of not more than 25 years.

This is an essential for ensuring the compliance of the legal regime regulating the development of natural resources in the moon and other celestial bodies with the non-appropriative nature of outer space sets out. The obtainments of the license will be subject to the payment of an initial fee.

The licensee will have also to pay a fee every five years. Additionally, in case he does not comply with certain terms of the license a fine can be imposed. The licensee enjoys two rights for the whole duration of the license: the rights of continued use over the area object of the license and the right to exercise property right over the extracted materials and the benefits generated thereof. Property rights over the extracted natural resources in the moon and other celestial bodies

They are necessary in order to provide the licensee with a reward for the effort they made to explore and exploit a lunar site and to make such exploitation a profitable business.

18.5.7. Exchange of Information

Members and the ISA shall facilitate the exchange of space policy, programs, scientific and technical information pertaining to the fields of space technology for the moon and other celestial bodies.

18.5.8. Education and Research

The ISA shall ensure the execution of basic activities, such as education (astronauts etc.), documentation, studies of future projects and technological research work. The ISA also facilitates the collection of relevant information and its dissemination to Member States, assistance and advice for harmonizing national and international programs and the elaboration and execution of scientific programs including the design, development, construction, launching, placing satellites and space shuttle in orbit and control of satellites and all similar activities for launching facilities, moon station, airport or space transport system from the earth to the moon and other celestial bodies.

18.5.9. International Cooperation

The ISA may, upon decisions of the Council taken by vote of a two-thirds majority of all Members States, cooperate with other international organizations and institutions and with Governments, organizations and institutions of non-Member States, and conclude agreements with them to this effect.

18.5.10. Financial Contributions

The ISA will be financed by its member States. The scale of contributions shall be based on the average national income of each Member State for the three latest years for which statistics are available.

18.5.11. Raising and Accumulation of Fund

According to Article 11, (7), (d) of the 1979 Moon Agreement, the ISA must raise and accumulate the funds for the equitable sharing by all States Parties in the benefits derived from those resources, whereby the interests and needs of the developing countries, as well as the efforts of those countries which have contributed either directly or indirectly to the exploration of the moon, shall be given special consideration.

18.5.12. Organization of ISA

An organization to be named the International Space Agency is formed by the Convention. It is made up of a General Assembly, a Council, a Director General, a Senior Staff and such other bodies as may be necessary.

18.5.12.1. The General Assembly

- The participation in the works of the General Assembly is, indeed, open to all States which have accepted the present legal regime and which are member of UNCOPUOS.
- The General Assembly, which is the principal organ, is composed of representatives of all Member States.
- The Assembly shall meet annually and shall be convened by the Council at a suitable time and place. Extraordinary meetings of the Assembly may be held at any time upon the call of the Council or at the request of any ten contracting States addressed to the Secretary General.
- It meets when it is required and is composed of either Ministers of the Member's States or government delegates. When it meets at ministerial level it can fulfill the political and practical function for the International conference on the exploitation and development of the moon and other celestial bodies.
- The General Assembly elects its Chairman and its Vice-Chairman for a period of three years, and re-election is possible for a further year.

- The Chairman shall direct the meetings, the proceedings, prepare the decisions and maintain appropriate contact with the Member States; he shall advice the Director General and obtain from him all necessary information.
- When the Council meets at the ministerial level, it shall elect a chairman for that meeting.

18.5.12.2. The Council

- The Council shall be a permanent body responsible to the Assembly. It shall be composed of twenty-four contracting States elected by the Assembly. An election shall be held at the first meeting of the Assembly and thereafter every two years, and the members of the Council so elected shall hold office until the next following election.
- The Council shall elect its President for a term of three years. He may be reelected. He shall have no vote.
- The Council shall elect from among its members one or more Vice Presidents who shall retain their right to vote when serving as acting President.
- The President need not be selected from among the representatives of the members the Council but, if a representative is elected, his seat shall be deemed vacant and it shall be filled by the State which he represented.
- Its functions and powers represent the core of the system including license.
- Regulating the exploitation of the mineral resources of the moon and Other Celestial Bodies.

18.5.12.3. Director General

- The Director General is the executive of the ISA and its representative.

The Director General, who is the head of the executive body, is appointed by Chairman of Board of Directors after two-thirds approval resolution is obtained from the Board of Directors.

- He is responsible for the management of the ISA, the execution of the programs and he accomplishes all the tasks imposed on him by the Council as well as the implementation of its policy and the attainment of its objectives in accordance with the ISA Convention.

18.5.12.4. Senior Staff

Members of senior staff for management, defined by the Council, shall be appointed by Chairman of the Council on the recommendation of the Director General.

18.5.13. Disputes and Arbitration

Disputes between Member States or between any of them and the ISA must first be settled by the Council. If the dispute is not settled this way, it shall at request of any party to the dispute be submitted to arbitration. Unless the Parties agree differently or the Council adopts other rules, the Arbitration Tribunal shall consist of three members of contracting states. Each Party shall appoint one of them, and those two arbitrators shall designate a third member. The third member is the umpire and presides over the tribunal.[37]The rules of procedure may be agreed between the Parties or imposed by the Council. The award shall be decided by the majority of votes (abstentions are not allowed) and it is final and binding.

18.5.14. Headquarter

The permanent seat of the ISA shall be determined by the resolution of the final meeting of UNCOPUOS and UN General Assembly. As the headquarters of the International Court of Justice and the International Criminal Court are located at The Hague (The Netherlands) and the headquarter of the International Tribunal for the Law of the Sea is located in Hamburg (Germany), already three International Court's headquarters are all located in Europe.

So, it has to be taken into account that Europe's 710 million people only make up 11% of the world's population but Asia accounts for over 60% of the world population with almost 3.8 billion people.[38] Therefore it should be adequate to decide on locating the ISA headquarters in the Asian region and since in Asia the Republic of Korea is located as a "buffer region" between China and Japan as the geopolitical powers, Seoul (the Republic of Korea), Beijing (China), Tokyo (Japan) or another city in the Asian Pacific region should be eligible for this purpose.

The location of the International Space Agency's headquarters is determined by a majority vote of the board of directors among Seoul, Beijing and Tokyo cities.

18.6. CONCLUSION

The International Space Agency (ISA) will be regarded as a new road for the global space policy and exploitation of the moon and other celestial bodies in the global community. The ISA also coordinates the broad thinking needed to meet new challenges in the global countries. The ISA will provide a bright prospects and vision of the global community's future in the moon and other celestial bodies, and for the benefits for mankind on the ground that satellites and spacecraft can supply.

As the moon, Mars, Asteroid, Saturn, Jupiter, Titan and other celestial bodies had been buried a large quantity of the precious natural resources, so we must establish a new International Space Agency in order to develop efficiently and effectively the aforementioned minerals. Due to the developments of internet, telecommunication by the satellites, spacecraft's and international space stations, it will be extinguished gradually or step by step the boundary among the developed countries. It is necessary for us to establish the ISA so as to work together in union, to strengthen cooperation in research, and to establish friendly relations for the benefit of the mankind.

Finally, a very important point is that a political drive, at the highest level, should be given to mobilize states to this initiative, possibly taking the form of a solemn statement by heads of the space superpower's countries setting out objectives and prospects for the long term. It should be noted that this political drive will be necessary not only to set up the organization, but also during a subsequent period.[39]

It is desirable for us to establish the ISA in order to develop and mine efficiently the natural resources of moon and other celestial bodies. I am sure that it is possible to establish the ISA in the near future, if the heads of the space super powers would agree to establish the ISA through a summit conference as well as conference of UNCOPUOS, General Assembly and Security Council of UN.

ADDITIONAL READING

Cordell, B. (1984). *A Preliminary Assessment of Martian Natural Resource Potential*. The Case for Mars II.

Namowitx, S., & Stone, D. (1975). *Earth Science: The World We Live In*. American Book Company.

Sorrell, C. (1973). *Rocks and Minerals*. Golden Press.

KEY TERMS AND DEFINITIONS

Asteroids: Asteroids are minor planets, especially of the inner Solar System. Larger asteroids have also been called planetoids. These terms have historically been applied to any astronomical object orbiting the Sun that did not resolve into a disc in a telescope and was not observed to have characteristics of an active comet such as a tail. As minor planets in the outer Solar System were discovered that were found to have volatile-rich surfaces similar to comets, these came to be distinguished from the objects found in the main asteroid belt. In this article, the term "asteroid" refers to the minor planets of the inner Solar System, including those co-orbital with Jupiter. Although their location in the asteroid belt excludes them from planet status, the three largest objects, Ceres (939km), Vesta (525km), and Pallas (512km), are intact protoplanet that share many characteristics common to planets, and are atypical compared to the majority of irregularly shaped asteroids. The fourth largest asteroid, Hygiea (434km), appears nearly spherical although it may have an undifferentiated interior, like the majority of asteroids. Between them, the four largest asteroids constitute half the mass of the asteroid belt.

Mars Crater: There are a number of different types of craters that have been observed and studied on Mars. Many of them are shaped by the effects of impacts into ice-rich ground. A study of the distribution of these craters demonstrated that the thickness of a frozen layer on Mars varies from about 1.3 km (equator) to 3.3 km (poles). This represents a great deal of frozen water. It would be equal to 200 meters of water spread over the entire planet, if one assumes the ground has 20% pore space. The researchers assumed that the single-layer ejecta craters would all be within the icy layer, but the double and multiple layer ejecta craters would always penetrate the icy layer. By finding an average of the largest single-layer ejecta crater depth and the smallest multiple-layer ejecta crater depth, the thickness of the icy layer, called the cryosphere was determined.

Mineral Rights: Mineral rights are property rights to exploit an area for the minerals it harbors. Mineral rights can be separate from property ownership (see Split estate). Mineral rights can refer to sedentary minerals that do not move below the Earth's surface or fluid minerals such as oil or natural gas. There are three major types of mineral property; unified estate, severed or split estate, and fractional ownership of minerals. Owning mineral rights (often referred to as a "mineral interest" or a "mineral estate") gives the owner the right to exploit, mine, and/or produce any or all minerals they own. Minerals can refer to oil, gas, coal, metal ores, stones, sands, or salts.

ENDNOTES

1 Fabio Tronchetti, *A legal regime to govern the exploitation of the natural resources of the Moon and other celestial bodies,* The Korean Journal of Air and Space Law (Vol.23,No.1, 2008), p.168.

2 https://en.wikipedia.org/wiki/Mars

3 https://en.wikipedia.org/wiki/Asteroid

4 https://en.wikipedia.org/wiki/Saturn

5 Jupiter is the fifth planet from the Sun and the largest in the Solar System.

6 https://en.wikipedia.org/wiki/Titan_(moon)

7 https://fti.neep.wisc.edu/fti?rm=gallery

8 See Sci./Tech. Moon map aids discovery at http://news.bbc.co.uk/1/hsci/tech/2260

9 https://en.wikipedia.org/wiki/Lunar_water

10 http://www.abc.net.au/news/newsitems/200411/s1252715.htm

11 https://www.space.com/25305-water-moon-earth-common-origin.html

12 https://spacenews.com/nasa-confirms-plans-to-send-prospecting-rover-to-the-moon

13 https://ja.wikipedia.org/wiki/%E7%81%AB%E6%98%9F

14 https://www.axismag.jp/posts/2018/12/111229.html

15 Cordell, B. 1984. A Preliminary Assessment of Martian Natural Resource Potential. The Case for Mars II.

16 Larry O'Hanlon (February 22, 2010). *"Mining Mars? Where's the Ore?"*. Discovery News.

17 Namowitx, S. and D. Stone. 1975. Earth Science: The World We Live In. American Book Company. NY, NY.

18 https://www.livescience.com/technology/Rare-Earth-Elements-100614.html

19 Sorrell, C. 1973. Rocks and Minerals. Golden Press. NY, NY.

20 http://link.springer.com/article/10.1007%2FBF02837988

21 https://www.google.com/search?source=hp&ei=CqJRXZfAOYqymAWsqJKgCg&q=havard

22 https://courrier.jp/news/archives/57948/?ate_cookie=1566819646

23 https://solarsystem.nasa.gov/planets/saturn/overview

24 https://www.spacetelescope.org/images/opo9828c

25 https://ja.wikipedia.org/wiki/%E5%9C%9F%E6%98%9F

26 https://ameblo.jp/yukikaze99j/entry-10075351005.html

27 https://www.unoosa.org/oosa/en/ourwork/copuos/lsc/2017/index.html

28 https://www.mofa.go.jp/mofaj/press/release/press4_007290.html

29 The Outer Space Treaty, formally known as the *Treaty on Principles Governing the Activities of States in the Exploration and Use of Outer Space, including the Moon and Other Celestial Bodies,* is a treaty that forms the basis of international space law. The treaty was opened for signature in the United States, the United Kingdom, and the Soviet Union on January 27, 1967, and entered into force on October 10, 1967. As of April 2018, 107 countries are parties to the treaty, while another 23 have signed the treaty but have not completed ratification.

30 *Agreement Governing the Activities of States on the Moon and Other Celestial Bodies*, (usually referred as Moon Agreement) signed on 18 December 1979; As of January 2018, only 18 states; Australia, Austria, Belgium, Chile, Kazakhstan, Lebanon, Mexico, Morocco, The Netherlands, Pakistan, Peru, Philippines, Saudi Arabia, Turkey and Uruguay, etc. have ratified it. France, Guatemala, India and Romania have signed but have not ratified it.

31 The refusal of the developed States to ratify the Moon Agreement was largely due to the insertion of the Common Heritage of Mankind idea in Article XI of the Agreement declaring the Moon and its resources to be "the Common Heritage of Mankind", see C.Q.Christol, "*Important concepts for international law of outer space*", in Proceedings of the Fortieth Colloquium on the Law of Outer Space, (1997), p. 73; F.G.von de Dunk, "*The dark side of the Moon: public concepts and private enterprises*", in Proceedings of the Fortieth Colloquium on the Law of Outer Space, (1997), p. 121.

32 Fabio Tronchetti, *op.cit.*, pp. 133~144.

33 Article 11, 5. States Parties to this Agreement hereby undertake to establish an international regime, including appropriate procedures, to govern the exploitation of the natural resources of the moon as such exploitation is about to become feasible. This provision shall be implemented in accordance with article 18 of this Agreement.

34 Article 18 Ten years after the entry into force of this Agreement, the question of the review of the Agreement shall be included in the provisional agenda of the General Assembly of the United Nations in order to consider, in the light of past application of the Agreement, whether it requires revision. However, at any time after the Agreement has been in force for five years, the Secretary-General of the United Nations, as depository, shall, at the request of one third of the States Parties to the Agreement and with the concurrence of the majority of the States Parties, convene a conference of the States Parties to review this Agreement. A review conference shall also consider the question of the implementation of the provisions of article 11, paragraph 5, on the basis of the principle referred to in paragraph 1 of that article and taking into account in particular any relevant technological developments.

35 In 1975, European Space Conference, meeting in Brussels, approved the text of the *"Convention for the Establishment of a European Space Agency"* setting up the European Space Agency. The member states are now fifteen countries: Austria, Belgium, Denmark, Finland, France, Germany, Ireland, Italy, the Netherlands, Norway, Portugal, Spain, Sweden, Switzerland, and the United Kingdom. Canada is a Cooperating State: United Nations, *"Space Activities of the United Nations and International Organizations"*, UN (New York, 1992), at 135; H.L. van Traa-Engelman, "*Commercial Utilization of Outer Space*," Martinus Nijhoff Publishers (1993), pp.160-161.

36 The Council will have two ways to verify if the licensee is operating in accordance with the license: 1) through a report which every license is obliged to provide on an annual basis containing information on the activities which have been undertaken; 2) through a manned mission which can check in loci the status of the exploitative activities. The Council should give the licensee one month notice before undertaking the control. The licensee shall offer proper collaboration and provide information during the control.

37 E. R. C. Van Bogaert, "*Aspects of Space Law*", Kluwer, (1986), at 271.

38 https://en.wikipedia.org/wiki/World_population

39 Gabriel Lafferranderie, *"Outlook on Space Law over the next 30 years",* Essays Published for the 30th Anniversary of the Outer Space Treaty, Kluwer Law International (1997), at 427.

Chapter 19
Proposal for Establishing an International Court of Air and Space Law

ABSTRACT

This chapter describes the proposal for establishing an "International Court of Air and Space Law (ICASL)." The establishment of an ICASL is necessary and should provide the uniformity of decisions currently lacking in the jurisdiction relating to disputes in the field of air and space law and space exploitation. It is desirable that the establishment of an ICASL should lead to the strengthening of international cooperation deemed essential by the global community towards the joint settlement of transnational claims and cases with respect to air and space law. In order to understand easily the global issues and solutions on the outer space law and policy, the author describes legal issues and prospects on outer space law and policy. The establishment of such a court should be designed to promote the speed of work and the fairness of trial in air and space law cases.

19.1. INTRODUCTION

The idea of establishing an International Court of Air and Space Law (hereinafter referred to as ICASL) is expressed as the personal opinion of the author in his capacity as an academician and a practitioner. The establishment of such a Court should be designed to promote the speed of work and the fairness of trial in air and space law cases. In the 21st century, the development of air and space science and technology will continue with ever greater rapidity. After half a century, some air and space superpowers and organizations such as the United States, the European Union, Russia, Japan and China have already scored remarkable achievements that greatly promoted the development of social productivity and progress. The continuous development and the application of air and space technology have obtained an important role in the modernization drive of the world community. The emergence of aerospace technology in the global countries has brought huge contributions to economic and social progress.

DOI: 10.4018/978-1-7998-7407-2.ch019

Nevertheless, in order to allow such an exploitation also in the future, a major problem must be solved, which is caused by the absence of a specific legal regime in air and space law providing specific and universally accepted rules on the future exploitation of the natural resources.

The establishment of an ICASL would lead to the strengthening of international cooperation deemed essential by the global community towards the joint settlement of transnational claims and cases with respect to air and space law.

It would also act as a catalyst for efforts[1] towards the solution of aircraft, satellite and space shuttle's accidents as well as cases relating to injures of crews and passengers.

Relevant information about such cases as well as the respective disputes and trials could then be centrally and independently managed for the benefit and in the interests of the global community.

For this purpose ICASL will be composed of a body of 14 judges, versed deeply in international air and space law who will be elected by a majority resolution of the Council and General Assembly of the International Civil Aviation Organization (ICAO) in cooperation with the Committee on the Peaceful Use of Outer Space (UNCOPUOS) and the UN General Assembly.

19.2. NECESSITY FOR THE CREATION OF INTERNATIONAL COURT OF AIR AND SPACE LAW

Reasons (in detail) why the ICASL should be established are the following.

It is a characteristic features of air aircraft and spacecraft accidents that ① great amounts of damage are involved and in some cases, ② operators suffer a total loss in the sense of "all or nothing", ③ accidents often involve catastrophes, and ④ there is a relation of subordination between the air or space assets and the ground, namely the relevant air traffic and space control systems in charge such as air traffic controller. In addition, ⑤ all accidents are of an international character.[2]

Therefore, the characteristics of air and space accidents are significantly different from road, railway and Maritime's accidents.

These aircraft, satellite and spacecraft's accidents have given cause for many disputes between the victims and the air and space carriers about questions of limited or unlimited liability and the appraisal of damages also due to sudden attacks, or accidents relating to the space shuttle or damage caused by space debris.

The establishment of ICASL could offer the unique opportunity unify the new rules Æiii(*Convention for the Unification of Certain Rules for International Carriage by Air*) of 1999 which are based on contractual liability, with the provisions of the new *Montreal Unlawful Interference Compensation Convention* (aircraft terror)[3] as well as the *General Risk Convention* of 2009[4] and the *Space Liability Convention* of 1972 which are based on tort liability.

Thereby a just and equitable solution could be found to disputes arising from disasters such as terror attacks of September 11, 2001[5] at New York and an orbit collision case between an US Iridium satellite and Russian Cosmos satellite of the 10th February, 2009.

As to the substantive and procedural law to be applied by ICASL, it must be uniform and conform to justice.

The establishment of other international courts such as the International Court of Justice (ICJ) based on the UN Charter of 1945[6], the International Tribunal for the Law of the Sea based on the United Nations Convention on the Law of the Sea, Annex of 1982 [7], the International Criminal Court[8] based on the Rome Statute of the International Criminal Court, the Court of the European Communities [9] at Luxembourg a and the European Human Rights Court[10] at Strasbourg demonstrates that the establishment of such an institution is possible. The ICASL will acts as a world court with global jurisdiction.[11] It will have dual jurisdiction: and will decides, in accordance with international law, disputes of a legal nature that will be submitted to it by States (jurisdiction in contentious cases). European Communities [12] at Luxembourg and the European Human Rights Court[13] at Strasbourg demonstrates that the establishment of such an institution is possible.

The ICASL will acts as a world court with global jurisdiction[14] It will have dual jurisdiction: and will decides, in accordance with international law, disputes of a legal nature that will be submitted to it by States (jurisdiction in contentious cases).

ICASL will gives advisory opinions on legal questions at the request of the organs of the United Nations or specialized agencies authorized to make such a request (advisory jurisdiction).[15]

The International Court of Justice has been accused in the past of moving at an unbearably slow pace and certain cases seem to bear this out: it took e.g. the court nearly 14 years (1993-2021) to get from the filing of application to a merits judgment in the *Bosnian Genocide Convention Case*, 11 years in the Oil Platforms Case, and 10 years in *Qatar v. Bahrain.*[16]

The alternative of ICASL would eliminate the lack of uniformity of decisions under the air and space conventions, protocols and agreements.[17] In addition, national courts would no longer have to apply their own choice of law analysis in choosing the applicable liability limits or even unlimited (absolute) liability which do not fall under such new air and space law system. Thus, the creation of ICASL would eliminate any disparity of damage awards among similarly situated parties of non-members States to different air and space conventions, protocols, agreements as well as national laws and jurisprudence.

However, the jurisdiction of ICASL should be further reaching than that of the International Court of Justice so that it can easier fulfill its mandate as explained.

As a first step a Draft Convention for on the Establishment of the ICASL must be drafted. In the following the main features of such a draft will be outlined in comparison with the Statute (Article 1~70) of the International Court of Justice [18], the Statue (Article 1~41) of the International Tribunal for the Law of the Sea[19] and the Statute (Article 1~128) of the International Criminal Court[20].

19.3. ORGANIZATION — COMPOSITION, ELECTION, PROCEDURE, DUTY, COMPETENCE, CHAMBERS, JURISDICTION, HEARING AND JUDGMENT OF THE ICASL

19.3.1. Composition of ICASL

The ICASL shall be the principal judicial organ of ICAO and UNCOPUOS. It shall be established and constituted by *"the Convention for the Creation of the International Court of Air and Space Law* (tentative title)" and shall function in accordance with the provisions of the said Convention and Statute.[21] The Court shall have international legal personality and have such legal capacity as may be necessary for the exercise of its functions and the fulfillment of its purposes.

According to its statue, the Court may also exercise its functions and powers on the territory of any State Party and, by special agreement, on the territory of any other State. The Court shall be composed of a body of independent judges, elected regardless of their nationality from among persons of high moral character, who possess the necessary qualifications in their respective countries for appointment to the highest judicial offices, or who are jurists and consultants of recognized competence in international air and space law.[22]

The International Court of Air and Space Law(ICASL) shall consist of fourteen judges (seven judges in the field of aviation law and seven judges in the field of space law), not two of whom may be nationals of the same state. No group may nominate more than four persons, not more than two of whom shall be of their own nationality. The Presidency is responsible for the proper administration of the court. It comprises the President and Vice-Presidents.

19.3.2. Procedure of Election for Judges of the Court

The judges for eight years term of the International Court of Air and Space Law shall be elected by the Council of International Civil Aviation Organization (ICAO) under the United Nations as well as General Assembly of the United Nations Committee on the Peaceful Use of Outer Space (UNCOPUOS) from a list of candidates nominated by the national groups from the seven continents and the aforementioned of two international organizations.

Before making these nominations, each national group is recommended to consult its highest court of justice, its legal faculties, schools of law as well as its academies and their respective sections devoted to the study of air and space law.[23]

The ICAO Secretary-General and the Director of the United Nations Office for Outer Space Affairs (UNOOSA) shall prepare a list in alphabetical order of all the persons thus nominated.

The list shall be submitted to the General Assembly of ICAO as well as to the UNCOPUOS and its Legal Subcommittee such as Secretary General of ICAO and Director of UNOOSA who shall proceed independently of one another in the election of the members of the Court by secret ballot. Those candidates who obtain an absolute majority of votes in the General Assembly of ICAO as well as in the UN General Assembly and in UNCOPUOS shall be considered as elected. In the event of more than one national of the same state obtaining an absolute majority of the votes both of the General Assembly of ICAO as well as from the UN General Assembly and of UNCOPUOS, the eldest of these only shall be considered as elected.

The Court shall elect its President and Vice-President for a four year's term but he may be re-elected. The Court shall appoint its Registrar and may provide for the appointment of such other officers as might be necessary.

19.3.3. Term, Duty, Diplomatic Privileges, Remuneration for Members of the Court

Though members of the Court shall be elected for eight years term, but he may be re-elected as one time. No member of the Court may exercise any political or administrative function or engage in any other occupation of a professional nature and associate actively with or be financially interested in any of the operations of any enterprise concerned with the exploration or exploitation of the resources of the air and outer space or other commercial uses of the air or outer space.

No member of the Court may participate in the decision of any case in which he has previously taken part as agent, counsel or advocate for one of the parties, or as a member of a national or international court or tribunal, or in any other capacity.[24] The members of the Court, when engaged in the business of the Court, shall enjoy diplomatic privileges and immunities. Every member of the Court shall, before taking up his duties, make a solemn declaration in open court that he will exercise his powers impartially and conscientiously.

This, however, shall not prevent the Court from sitting and exercising its functions elsewhere whenever the Court considers it desirable. The President and the Registrar shall reside at the seat of the Court. Each elected member of the Court shall receive an annual allowance and, for each day on which he exercises his functions, a special allowance provided that in any year the total sum payable to any member as special allowance will not exceed the amount of the annual allowance.[25]

The President shall receive a special annual allowance. The Vice-President shall receive a special allowance for each day on which he acts as President. These salaries and allowances shall be decided by the General Assembly of ICAO and by UNCOPUOS taking into account the workload of the Court. They may not be decreased during the term of office.

19.3.4. Challenge, Formation for Chambers of the Court

If, for some special reason, a member of the Court considers that he should not take part in the decision of a particular case, he shall so inform the President. If the President considers that for some special reason one of the members of the Court should not sit in a particular case, he shall give him notice accordingly. If in any such case the member Court and the President disagree, the matter shall be settled by the decision of the Court.[26]

The Court may from time to time form one or more chambers, composed of three or more judges as the Court may determine, for dealing with particular categories of cases; for example, cases of aircraft hijacking, compensation for damage caused by aircraft accidents, accidents due to the launching of satellites as well as spacecraft and cases relating to the collision among aircraft, spacecraft and satellites as well as damage caused by and space debris etc.

The Court may at any time also form a chamber for dealing with a particular case. The number of judges to constitute such a chamber shall be determined by the Court with the approval of the parties. The Court shall frame rules for carrying out these functions. In particular, it shall lay down rules of procedure.

The decisions of the Chamber shall be enforceable in the territories of the States Parties in the same manner as judgments or orders of the highest court of the State Party in whose territory the enforcement is sought.

19.3.5. Competence, Jurisdiction of the Court and Nationality of Members

The Court shall be open to States Parties. Only states may be parties in cases before the Court. The Court, subject to and in conformity with its Rules, may request of public international organizations information relevant to cases before it, and shall receive such information presented by such organizations on their own initiative. The Court shall be open to the states parties to the Statute.

The jurisdiction of the Court comprises all cases which the parties refer to it and all matters specially provided for in the air and space treaties and conventions in force.

In the event of a dispute as to whether the Court has jurisdiction, the matter shall be settled by the decision of the Court.

The Court shall decide in accordance as the following items.

1. relevant international conventions, whether general or particular, which were ratified or acceded to by the contesting states; international custom, as evidence of a general practice accepted as law;
2. general principles of law recognized by civilized nations;

However this does not prejudice the power of the Court to decide a case (Latin term for "according to the right and good" or "from equity and conscience"), if the parties agree thereto.[27]

Members of the Court of the nationality of any of the parties to a dispute shall retain their right to participate as members of the Court. If the Court, when hearing a dispute, includes upon the bench a member of the nationality of one of the parties, any other party may choose a person to participate as a member of the Court. If the Court, when hearing a dispute, does not include upon the bench a member of the nationality of the parties, each of those parties may choose a person to participate as a member of the Court.

19.3.6. Procedure, Hearing, Expenses of the Court

Disputes are submitted to the Court, as the case may be, either by notification of a special agreement or by written application, addressed to the Registrar. In either case, the subject of the dispute and the parties shall be indicated. The Registrar shall forthwith notify the special agreement or the application to all concerned.

The Registrar shall also notify all States Parties. The Court shall make orders for the conduct of the case, decide the form and time in which each party must conclude its arguments, and make all arrangements connected with the taking of evidence.[28]The Court shall have the power to indicate, if it considers that circumstances so require, any provisional measures which ought to be taken to preserve the respective rights of either party.

1. The parties shall be represented by agents.
2. They may have the assistance of counsel or advocates before the Court.
3. The agents, counsel, and advocates of parties before the Court shall enjoy the privileges and immunities necessary to the independent exercise of their duties.

The procedure shall consist of two parts: written and oral. The hearing shall be under the control of the President or, if he is unable to preside, of the Vice-President; if neither is able to preside, the senior judge present shall preside.

The hearing in Court shall be public, unless the Court shall decide otherwise, or unless the parties demand that the public be not admitted. Minutes shall be made at each hearing and signed by the Registrar and the President.

The Court shall make orders for the conduct of the case, shall decide the form and time in which each party must conclude its arguments, and make all arrangements connected with the taking of evidence. Whenever one of the parties does not appear before the Court, or fails to defend its case, the other party may call upon the Court to decide in favor of its claim.

When, subject to the control of the Court, the agents, counsel, and advocates have completed their presentation of the case, the President shall declare the hearing closed. All questions shall be decided by a majority of the judges present. In the event of an equality of votes, the President or the judge who acts in his place shall have a casting vote. The expenses of the Court shall be borne by the States Parties and the Authority on such terms and in such a manner as shall be decided at meetings of the States Parties. When an entity other than a State Party or the Authority is a party to a case submitted to it, the Court shall fix the amount which that party is to contribute towards the expenses of the Court. Unless otherwise decided by the Court, each party shall bear its own costs.

19.3.7. Judgment, Advisory Opinions, Finality and Binding Force of Decisions

The judgment shall state the reasons on which it is based. It shall contain the names of the judges who have taken part in the decision. It shall be signed by the President and by the Registrar. It shall be read in open court, due notice having been given to the agents. The Court may give an advisory opinion on any legal question at the request of whatever body may be authorized by or in accordance with the Statue of the ICASL to make such a request.[29]

The Registrar shall forthwith give notice of the request for an advisory opinion to all states entitled to appear before the Court. The decision of the Court is final and shall be complied with by all the parties to the dispute. The decision shall have no binding force except between the parties in respect of that particular dispute.

In the event of a dispute as to the meaning or scope of the decision, the Court shall construe it upon the request of any party. The official languages of the Court shall be French and English. If the parties agree that the case shall be conducted in English, the judgment shall be delivered in English.

19.3.8. Headquarter

The permanent seat of the ICASL shall be determined by the resolution of the final meeting of ICAO, UNCOPUOS and UN General Assembly. As the headquarters of the International Court of Justice and the International Criminal Court are located at The Hague (The Netherlands) and the headquarter of the International Tribunal for the Law of the Sea is located in Hamburg (Germany), already three International Court's headquarters are all located in Europe.

So, it has to be taken into account that Europe's 710 million people only make up 11% of the world's population but Asia accounts for over 60% of the world population with almost 3.8 billion people.[30] Therefore it should be adequate to decide on locating the ISA headquarters in the Asian region and since in Asia the Republic of Korea is located as a "buffer region" between China and Japan as the geopolitical powers, Seoul (the Republic of Korea), Beijing (China), Tokyo (Japan) or another city in the Asian Pacific region should be eligible for this purpose.

19.4. CONCLUSION

If the International Civil Air Organization (ICAO) and UN Committee on the Peaceful Use of Outer Space (UNCOPUOS) will be established ICASL in future, it would lead to the strengthening of international cooperation deemed essential by the global community towards the joint settlement of transnational claims and cases in the field of the international air and space law.

The establishment of an International Court of Air and Space Law is necessary and should provide the uniformity of decisions currently lacking in the jurisdiction relating to disputes in the field of air and space law and space exploitation.

In order to achieve this aim a political drive, at the highest level is necessary to motivate and mobilize states toward this aim, which will possibly have to take the form of a solemn statement by heads States which are super powers setting out objectives and prospects for the long term.

If the summit conference among the world's great powers agree to the establishment of the International Aerospace Court in the future, I am confident that it will be possible.

ADDITIONAL READING

Kolb, R. (2019). *The International Court of Justice*. Hart Publishing.

Murphy, S. D. (2011). *The International Court of Justice*. Elsevier.

Vaughan & Fitzmaurice. (2009). *Fifty Years of the International Court of Justice*. Cambridge University Press. https://en.wikipedia.org/wiki/September_11_attacks https://www.ejiltalk.org/recent-developments-at-the-international-court-of-justice

KEY TERMS AND DEFINITIONS

COPUOS: The United Nations Committee on the Peaceful Uses of Outer Space (COPUOS) was established in 1958 (one year after the launch of Sputnik) as an *ad hoc* committee of the United Nations (UN). In 1959, it was formally established as a permanent body by United Nations resolution 1472 (XIV). All documents related to the Committee and its subcommittees, the Scientific and Technical Subcommittee and the Legal Subcommittee, are published on the UNOOSA website. The Committee was first established by the General Assembly in its resolution 1348 (XIII) of 13 December 1958 and was originally composed of 18 members. It has grown to include 92 members as of 2019, and is subsequently one of the largest committees of the General Assembly of the United Nations.

Court of the European Communities: The Court of Justice of the European Union (CJEU) is the judicial branch of the European Union (EU). Seated in the Kirchberg quarter of Luxembourg City, Luxembourg, this EU institution consists of two separate courts: the Court of Justice and the General Court. From 2005 to 2016 it also consisted of the Civil Service Tribunal. It has a *sui generis* court system, meaning 'of its own kind', and is a supranational institution. CJEU is the chief judicial authority of the European Union and oversees the uniform application. and CJEU also resolves legal disputes between national governments and EU institutions.

International Criminal Court: The International Criminal Court (ICC) is an intergovernmental organization and international tribunal that sits in The Hague, Netherlands. The ICC is the first and only permanent international court with jurisdiction to prosecute individuals for the international crimes of genocide, crimes against humanity, war crimes, and the crime of aggression. It is intended to complement existing national judicial systems and it may therefore exercise its jurisdiction only when national courts are unwilling or unable to prosecute criminals.

ENDNOTES

1 Gabriel Lafferranderie, *"Outlook on Space Law over the Next 30 years"*, Kluwer Law International, 1997, at 424.

2 Doo Hwan Kim, "*A Study on the Civil Liability of the Air Carriers and Legislative Problems*", Doctoral Dissertation (1984), Graduate School, Kyonghee University at Seoul in Korea, at 1.
3 Convention on Compensation for Damage to Third Parties, Resulting from Acts of Unlawful Interference Involving Aircraft (Unlawful Interference Convention) of 2009.

4 Convention on Compensation for Damage Caused by Aircraft to Third Parties

5 https://en.wikipedia.org/wiki/September_11_attacks

6 http://www.icj-cij.org/icjwww/igeneralinformation/inotice.pdf; The International Court of Justice was created by the Article 92~96 of the United Nations Charter in 1945 and is also a judicial principal organization of the United Nations. The court handles disputes between nations and it begun work in 1946. Its seat is at the Peace Palace at The Hague in The Netherlands.

7 https://www.itlos.org/start2_en.html; The International Tribunal for the Law of the Sea is an independent judicial body established by the Convention to adjudicate disputes arising out of the interpretation and application of the Convention. The Tribunal is composed of 21 independent members, elected from among persons enjoying the highest reputation for fairness and integrity and of recognized competence in the field of the law of the sea. The United Nations Convention on the Law of the Sea was opened for signature at Montego Bay, Jamaica, on 10 December 1982. It entered into force 12 years later, on 16 November 1994. A subsequent Agreement relating to the implementation of Part XI of the Convention was adopted on 28 July 1994 and entered into force on 28 July 1996. This Agreement and Part XI of the Convention are to be interpreted and applied together as a single instrument.

8 The Rome Statute of the International Criminal Court (often referred to as the International Criminal Court Statute or the Rome Statute) is the treaty that established the International Criminal Court (ICC). It was adopted at a diplomatic conference in Rome on 17 July 1998 and it entered into force on 1 July 2002. As of June 2009, 108 states are party to the statute. Chile will become the 109th party on 1 September 2009, and a further 39 states have signed but not ratified the treaty. Among other things, the statute establishes the court's functions, jurisdiction and structure. The International Criminal Court (ICC) is an independent, permanent court that tries persons accused of the most serious crimes of international concern, namely genocide, crimes against humanity and war crimes; Rome Statute of the Internationalinal_Court) Comment, "*The Role of Choice of Law in Determining Damage for International Aviation Accidents,*" 51 J. Air L. & Com. 953, 957 n. 301 (1986); *Id.* For a list of seventy cases filed with the court as of July, 1984, see 1983-1984 I.C.J.Y.B., 1984, at 3-6.

9 The European Court of Justice (ECJ) (officially the Court of Justice of the European Communities), is the highest court in the European Union in matters of European Union law. It is tasked with interpreting EU law and ensuring its equal application across all EU member states. The Court was established in 1952 and is based in Luxembourg. It is composed of one judge per member state —— currently 27 - although it normally hears cases in panels of three, five or thirteen judges. The court is led by a President who has been Vassilios Skouris since 2003; https://en.wikipedia.org/wiki/European_Court_of_Justice

10 The European Court of Human Rights (ECtHR) (French: Cour europeane des droits de l'homme) in Strasbourg is an international judicial body established under the European Convention on Human Rights (ECHR) of 1950 to monitor respect of human rights by states. The European Convention on Human Rights, or formally named Convention for the Protection of Human Rights and Fundamental Freedoms, is a convention adopted by the Council of Europe. All 47 member states of the Council of Europe are parties to the Convention. Applications against Contracting Parties for human rights violations can be brought before the Court by other states, other parties or individuals.

11 V.S. Verschehetin, "*The International Court of Justice as Potential Forum for the Resolution of Space Law Disputes*,"Luft-und Weltraumrecht im 21. Jahrhundert (Air and Space Law in the 21st Century), Liber Amicorum, Karl-Heinz Böckstiegel, edited by Marietta Benkö and Walter Kröll, Carl Heymanns Verlag KG, 2001, at 476.

12 The European Court of Justice (ECJ) (officially the Court of Justice of the European Communities), is the highest court in the European Union in matters of European Union law. It is tasked with interpreting EU law and ensuring its equal application across all EU member states. The Court was established in 1952 and is based in Luxembourg. It is composed of one judge per member state —— currently 27 - although it normally hears cases in panels of three, five or thirteen judges. The court is led by a President who has been Vassilios Skouris since 2003; https://en.wikipedia.org/wiki/European_Court_of_Justice

13 The European Court of Human Rights (ECtHR) (French: Cour europenne des droits de l'homme) in Strasbourg is an international judicial body established under the European Convention on Human Rights (ECHR) of 1950 to monitor respect of human rights by states. The European Convention on Human Rights, or formally named Convention for the Protection of Human Rights and Fundamental Freedoms, is a convention adopted by the Council of Europe. All 47 member states of the Council of Europe are parties to the Convention. Applications against Contracting Parties for human rights violations can be brought before the Court by other states, other parties or individuals.

14 V.S. Verschehetin, "*The International Court of Justice as Potential Forum for the Resolution of Space Law Disputes*, "Luft-und Weltraumrecht im 21. Jahrhundert (Air and Space Law in the 21st Century), Liber Amicorum, Karl-Heinz Böckstiegel, edited by Marietta Benkö and Walter Kröll, Carl Heymanns Verlag KG, 2001, at 476.

15 http://www.icj-cij.org/jurisdiction/index.php?p1=5

16 https://www.ejiltalk.org/recent-developments-at-the-international-court-of-justice/

17 The Warsaw system need not be destroyed, however. Professor Matte of McGill University suggests creating an International Court of Appeals or extending jurisdiction to the International Court of Justice at the Hague to decide Warsaw Convention cases.

18 http://www.icj-cij.org/documents/index.php?p1=4&p2=2&p3=0

19 https://www.itlos.org/documents_publications/documents/statute_en.pdf

20 https://untreaty.un.org/cod/icc/statute/romefra.htm

21 Articles of Incorporation, Memorandum of Association

22 Article 2, Statute of the International Court of Justice.

23 Article 6, Statute of the International Court of Justice.

24 Article 8 (Conditions relating to participation of members in a particular case), the Statute of the International Tribunal for the Law of the Sea.

25 Article 18 (Remuneration of members), the Statute of the International Tribunal for the Law of the Sea.

26 http://www.icj-cij.org/documents/index.php?p1=4&p2=2&p3=0

27 Article 36, Statute of the International Court of Justice.

28 Article 27 (Conduct of case), the Statute of the International Tribunal for the Law of the Sea.

29 Article 36, Statute of the International Court of Justice.

30 https://en.wikipedia.org/wiki/World_population

About the Author

Doo Hwan Kim lives in Seoul, the Republic of Korea. Author is now a Honorary President, Korea Society of Air & Space Law and Policy, Seoul, 1999~present. He is now also Visiting Professor of Law School, Beijing Institute of Technology (2000 –present) and Tianjin University Law School (2018-present) in China. He retired in 1999 as Professor in Law College, Soongsil University, Seoul, Korea after appointed in 1981. He holds LL.B from Law College, Seoul National University in 1957 and LL.M from Law Department, Graduate School, Seoul National University in Seoul, Korea in 1959. He also hold JSD (Jurisprudence Science Doctor) from Law Department, Graduate School, Kyung Hee University in Seoul, Korea in 1984. He was a Visiting Scholar of Law School, University of California Los Angeles (UCLA), Washington College of Law, The American University USA and Institute of Air and Space Law, McGill University (Montreal, Canada) 1990-91. He served as Dean of the Law College of Soongsil University 1986-1988, 1992-1996. He was also Honorary Professor of Gujarat National Law University in India, 2004- 2010. He was a President of the Korean Association of Air and Space Law 1993-1999. He served at the Advisory Council on Policy-Making of Prime Minister as a Legal Advisor, 1980-1983, Advisory Council on Policy-Making of Ministry of Transportation as a legal Advisor 1980-1985 and Advisory Council on Policy-Making of Ministry of Justice of the Korean Government 1985-1989. His Biography published at the Asia/Pacific-Who's-Who (India) 1993-present. Who's Who in the World (USA) 1999-present. He received a "*National Magnolia Gold Medal*" from the Korean Government, an "Outstanding Contribution Award" from Space Law Institute of Beijing Institute of Technology, Beijing, China, 2016 and a "Global Achievers Award" from the Indian Institute of Oriental Heritage in Kolkata, India on Feb., 2020. He was published one books in the field of "Air and Space Policy & Law" by Kluwer Law International Publisher in The Hague, Netherlands, 2000 and two books in the field of "Air and Space Policy & Law" by the Korean Publishers in Seoul, Korea, 2005-2011. His 62 papers written by English, Chinese and Japanese language in the field of air and space law and policy was published to the famous law Journal of USA, Canada, Germany, China, Japan, Japan, India etc. 1992-2020 and domestic 119 papers written by the Korean language was published in Seoul, Korea, 1980-2020. His total paper is 181 papers.

Index

A

B

C

E

H

I

J

K

L

M

N

O

S

U

V

W

Ensure Quality Research is Introduced to the Academic Community

Become an Evaluator for IGI Global Authored Book Projects

The overall success of an authored book project is dependent on quality and timely manuscript evaluations.

Applications and Inquiries may be sent to:
development@igi-global.com

Applicants must have a doctorate (or equivalent degree) as well as publishing, research, and reviewing experience. Authored Book Evaluators are appointed for one-year terms and are expected to complete at least three evaluations per term. Upon successful completion of this term, evaluators can be considered for an additional term.

If you have a colleague that may be interested in this opportunity, we encourage you to share this information with them.

www.ingramcontent.com/pod-product-compliance
Lightning Source LLC
LaVergne TN
LVHW060250011026
849161LV00011B/2684
* 9 7 8 1 7 9 9 8 7 4 0 7 2 *